D1049709

Global and Regional Changes in Atmospheric Composition

Ernö Mészáros
Department of Analytical Chemistry
Veszprém University
Veszprém, Hungary

Library of Congress Cataloging-in-Publication Data

Global and regional changes in atmospheric composition /
by Ernö Mészáros
p. cm.
Includes bibliographical references and index.
ISBN 0-87371-662-0
1. Atmospheric chemistry. 2. Atmosphere, Upper.
3. Man-Influence on nature. 4. Biosphere.
5. Air—pollution—Meteorological aspects. I. Title.
TD426.M62 1993
628.5′5—dc20

92-16369
CIP

Direct all inquiries to CRC Press, Inc., 2000 Corporate Blvd., N.W., Boca Raton, Florida 33431.

PRINTED IN THE UNITED STATES OF AMERICA
1 2 3 4 5 6 7 8 9 0
Printed on acid-free paper

The atmosphere is the face of the planet.
J.P. Lovelock, *The Ages of Gaia*

Ernö Mészáros was born in Budapest, Hungary in 1935. He finished his studies in 1957 at the University L. Eötvös (Budapest) receiving a diploma in meteorology. He obtained his doctorate degree in physics in 1961 at the same university. In 1985 he became a member of the Hungarian Academy of Sciences. He is presently the president of the Department of Earth Sciences of the Academy. At the same time he is the doctor "honoris causa" of the Université de Bretagne Occidentale (Brittany, France) and a member of the "Academia Europaea."

For eight years he served as the secretary of the Commission on Cloud Physics of the International Association of Meteorology and Atmospheric Physics (IUGG) and as a member of the Commission of Atmospheric Chemistry and Global Air Pollution of the same Association. He has taken an active part in the background air pollution monitoring program of the World Meteorological Organization, mainly as the principal director of its training center in Budapest.

Dr. Mészáros held research positions in the Meteorological Service of Hungary and he was director of the Institute for Atmospheric Physics of the Service between 1976 and 1990. He has published more than one hundred papers and several books on atmospheric physics and chemistry, including his volume on *Atmospheric Chemistry* in 1981.

Since 1992 he has been a professor of environmental sciences at the University of Veszprém, Hungary.

Preface

The Earth's atmosphere has a unique composition compared to the gaseous cover of other planets and moons in the solar system. It has developed in a close relationship with the biosphere, and its composition has been a consequence of the evolution of life. This means that the composition has changed parallel with changes in the biosphere. At the same time, the composition of the atmosphere has influenced the environment of living species. Since present composition and climate control human life in a significant way, the study of their possible future changes is of crucial importance.

This question is of particular interest in our days; human activities have become an essential factor in the control of atmospheric composition owing to the release of different pollutants into the atmosphere. It is now evident that urban air pollution affects human health, regional and continental air pollution cause crop losses and forest damages, and long-lived pollutants produce problems like global warming and ozone holes.

The aim of this book is to summarize regional and global changes in atmospheric composition induced by human activities together with present and future consequences of such modifications. The author wanted to write a coherent text, understandable for students in meteorology, chemistry, and ecology and beginners in the subject. He hopes that policy makers responsible for regional and global air quality management and regulation will also find the volume useful. It follows from this approach that it is not intended to present a complete literature review of this rapidly growing field. It is believed, however, that the references cited give a good overview of changes in atmospheric composition and they make further, more detailed studies for interested readers possible.

The author is indebted to Ms. A. Molnár for many fascinating discussions during the writing of the text and to his secretary, Ms. M. Antal for the careful preparation of the manuscript.

E. Mészáros

Contents

1

Introduction: Past Changes in Atmospheric Environment

1.1 GENERAL INTRODUCTORY REMARKS

Several billion years ago special terrestrial conditions made the formation of life possible. Since its formation the biosphere has played an active role in the control of environmental conditions. An interaction has developed between the evolution of living species and the environment.

During geological times the environment has changed in many respects. Changes in environmental conditions modified the biosphere and vice versa. However, environmental changes have not been relatively drastic during the past 3.9 billion years and have never endangered life as a whole. Even such events like the catastrophe of 60 million years ago, which led to the extinction of dinosaurs, created favorable conditions for the development of another part of animal species—mammals.

When life began the Earth's atmosphere was not similar to the present air. It was practically free of oxygen and contained, besides nitrogen, carbon dioxide like the atmospheres of Venus and Mars at present. The main peculiarity of our atmosphere—the presence of oxygen—is the result of biospheric evolution.

The climate of our planet also varied in the past: "ice ages" were followed by interglacial periods. In interaction with climate variations the composition of the atmosphere also changed.

Natural changes in atmospheric composition and climate were slow processes relative to human time scales. A glaciation period or ice age built up over a time of nearly 100,000 years and terminated in about 10,000 years. Thus, the last ice age reached its maximum 18,000 years ago. During the last 8,000 to 10,000 years the climate has been stable, with a relatively constant average temperature. This means that during our history the atmospheric environment has not changed. The stability has been favorable for humans and made social and economic development possible.

In the present industrial era this development has reached such a level that human activities have become able to modify environmental conditions on a time scale (~100 years) that is short relative to the period of natural changes. Modification of the air composition due to the release of different pollutants can lead to serious global problems like changes in our favorable climate.

The climate-composition relationship can be illustrated by the interaction of solar and terrestrial radiations and the atmosphere. The energy reaching our planet comes from the sun in the form of shortwave radiation. About a third of this radiation (called the planetary albedo) is scattered back to space mostly by clouds. Another 20% is absorbed by atmospheric constituents like water vapor, ozone, and aerosol particles. The balance warms up the Earth's surface. Owing to its temperature, the surface emits longwave radiation. A large part of this heat radiation is absorbed by atmospheric gases and clouds and reemitted downwards. Since chemical species (CO_2, CH_4, N_2O) absorbing longwave radiation are transparent to visible solar radiation, they are termed *greenhouse gases*. Without the presence of greenhouse gases and clouds the mean temperature at the surface would be as low as –18°C.

The difference in energy received and emitted by a given area determines the temperature, while its spatial distribution operates the atmospheric wind system called the *general circulation*. This indicates that the composition even controls, in some measure, the dynamics of the atmosphere.

On the other hand, the composition of the atmosphere is a function of the huge material flow of different elements and compounds in nature called the *biogeochemical cycles*. These cycles transport substances from one natural reservoir (atmosphere, hydrosphere, pedosphere, lithosphere, and biosphere) to another. The composition of a reservoir is constant if the material input is equal to the output; the sources are balanced by the sinks. Due to the interaction of sources and sinks each element/compound resides in a reservoir during a certain time termed the *residence time.* The residence time of greenhouse gases in the atmosphere is much longer than the characteristic transport time in the troposphere (the lowest: 10 to 15 km). For this reason their mixing is practically complete; greenhouse gases constitute the class of global air pollutants. Global air pollutants are the main agents for possible man-made climatic changes. In addition to this, some anthropogenic greenhouse gases like chlorofluorocarbons (freons) take part in the ozone destruction in the air layer between 15 and 55 km (the stratosphere).

There are many other gaseous pollutants (e.g., sulfur and nitrogen oxides) that are easily oxidized in the atmosphere to form acid vapors. These vapors condense in the air with water molecules to create minute droplets, called aerosols. Acidic gases and aerosol particles are removed from the air by dry and wet deposition. While dry deposition is controlled by turbulent diffusion in the air, wet deposition is a function of cloud and precipitation formation, and its magnitude is determined by the amount of rain and snowfall. Due to removal processes the residence time of these species is shorter (some days or week) than that of greenhouse gases. In interaction with ammonia they produce such regional problems as the acidification of the soil, lakes, rivers, and surface waters.

Human activities modify not only the acidity of deposition. Deposition is also altered by the anthropogenic release of different metals to the atmosphere. While the emissions in absolute terms are very low, metals of anthropogenic origin increase the amount of micronutrients in biological systems. Their accumulation on a large scale in vegetation and animal bodies increases the microelement content of foods, which can be toxic for human health.

The oxidation in the atmosphere is caused by the presence of oxygen containing free radicals. In the surface air, for the formation of these electrically uncharged, very reactive atomic groups, ozone molecules are necessary. Since the formation of ozone in the troposphere is controlled by the concentration of nitrogen oxides and organic compounds emitted partly by anthropogenic sources, man is also able to modify the oxidizing state of the air. Ozone and free radical formation is promoted by solar radiation, which means that photochemical reactions are involved in the process. For this reason photochemical "smog" formation and acidification are in close relationship.

The aim of this book is to present and discuss future global and regional anthropogenic changes in atmospheric composition as well as their environmental consequences. However, this is possible only when atmospheric pathways of elements/compounds of natural origin are also considered. In this way the magnitude of man-made sources can be compared to the release of natural species. Moreover, the impact of anthropogenic compounds on different phenomena can be identified. It follows from this approach that the scope of this book is closely related to a new and prosperous branch of atmospheric sciences termed the *air chemistry*. It is evident that before starting the discussion of present and future changes, the natural variations of atmospheric composition occurred in the past must also be briefly presented.

1.2 EVOLUTION OF THE EARTH'S ATMOSPHERE

1.2.1 The Prebiologic Atmosphere

Our knowledge of the composition of the atmosphere covering the Earth about 4.5 billion years ago is rather speculative. The abundance of rare gases on Earth compared to their concentration in the solar nebula (see, e.g., Warneck, 1988) shows that the early atmosphere consisted of volatile gases exhaled from the Earth's interior. As evidenced by very old minerals, at that time iron in the upper part of the mantle was in a reduced state. Consequently the composition of the earliest volcanic gases was somewhere between their present composition and the composition in equilibrium with metallic iron (Holland, 1984). Volcanic gases today are composed mainly of H_2O and CO_2, but they contain a small amount of SO_2 and N_2. In volcanic gases in a reduced environment, about the half of hydrogen is in molecular form (H_2), CO is the most abundant carbon species, and H_2S is the dominant sulfur compound. As in the previous case N_2 is the main nitrogen containing gas.

Results of photochemical model calculations (Kasting et al., 1979) show that in such a gas mixture the oxygen concentra-

Table 1.1 The Reactions Leading to Oxygen Formation in the Prebiotic Atmosphere[a]

H_2O	+	hv	→	H	+	OH	
H_2O	+	hv	→	H_2	+	O	
CO	+	OH	→	CO_2	+	H	
CO_2	+	hv	→	CO	+	O	
OH	+	OH	→	H_2O	+	O	
O	+	O + M	→	O_2	+	M	z = 60 km
O	+	OH	→	O_2	+	H	z < 30 km

[a] hv represents the photons coming from the sun, while z is the height and M is a third body.

tion depended on the hydrogen and carbon monoxide emission rate, as well as on the rate of oxidation of the Earth's crust. The water molecules were partly dissociated under the effects of strong ultraviolet radiation coming from the sun. In this way atomic and molecular hydrogen as well as oxygen atoms and hydroxyl free radicals (OH) were formed. While hydrogen escaped to space, OH radicals oxidized CO molecules into carbon dioxide. CO_2 was also photodissociated, which resulted again in atomic oxygen formation. This latter process gave the majority of oxygen atoms. On the other hand, the reaction of OH itself provided a smaller oxygen atom source. Molecular oxygen was formed either by the combination of oxygen atoms at an altitude of 60 km or by the reaction of atomic oxygen with hydroxyl radicals below 30 km (see Table 1.1).

Assuming conditions appropriate to this early time Kasting et al. (1979) calculate that the number of O_2 molecules was about 10^7 cm^{-3} at the surface* and it reached a magnitude of 10^{11} cm^{-3} at an altitude of 40 km. The hydrogen number density was around 10^{11} cm^{-3}, but at all heights the composition was dominated by CO_2 and mainly N_2 molecules. This was due to the fact that N_2 is an inert gas, and it accumulated in the atmosphere in spite of its low concentration in volcanic gases.

1.2.2 The Rise and Budget of Oxygen

The main reason for the presence of life and the pressure of oxygen on the Earth is the accumulation of water on the planet at very early times. Life was formed in the upper layers of

* The present value is 5.629×10^{18} cm^{-3}.

liquid water under the influence of ultraviolet solar radiation. Some earlier experiments (Miller, 1953) suggest that the first complex molecules were synthesized in water in contact with an atmosphere containing methane and ammonia. However, more recent kinetic calculations indicate that these gases were present only as trace constituents, that is, life was formed under less reducing conditions. According to our previous discussion, the primitive atmosphere consisted of N_2, CO/CO_2, and H_2/H_2O. It is very probable that, in the atmosphere, carbon monoxide and carbon dioxide were transformed into formaldehyde, a small but important fraction of which was removed by precipitation and delivered to the oceans (Holland, 1984). This process played an essential role in the synthesis of prebiologic organic molecules.

The first bacteria (the prokaryotes) used hydrogen or hydrogen sulfide to produce organics from carbon dioxide (Walker, 1974; Holland, 1984; Warneck, 1988). While the direct H_2 uptake is exoergic, the use of H_2S requires an energy input. The appearance of cyanobacteria (blue-green algae) about 3.5 billion years ago made a revolution in biospheric and atmospheric evolution. These bacteria can build organics from CO_2 and H_2S, but under suitable conditions they use water as a hydrogen donor. This form of photosynthesis, which has become the basic life process of plants, produced molecular oxygen from water molecules:

$$CO_2 + 2H_2O + h\nu \underset{\text{respiration/decay}}{\overset{\text{photosynthesis}}{\rightleftarrows}} CH_2O + H_2O + O_2$$

where the symbol hν represents the energy of solar radiation (photons). The oxygen emitted was a poison for a certain part of the early biosphere, but the oxidation of iron, hydrogen sulfide, and other reduced species held the oxygen concentration at a low level.

About 2.5 billion years ago, before present reduced compounds available were oxidized and the oxygen concentration in the atmosphere and ocean began to rise. 2.0 billion years ago the oxygen concentration was 0.1% of the present level (Schidlowski, 1978) and it reached 1% around 1.4 billion years ago. In this period new living species (the eukaryotes) appeared in the oxidizing ocean, which used oxygen and organic matter to produce energy (respiration). Even after some time these new organisms learned to eat not only organic debris but also the photosynhetizers.

When the atmospheric oxygen level was equal to 10% (420 million years ago) an effective ozone layer came into being (Ratner and Walker, 1972) to shield the surface from lethal ultraviolet radiation.* This was the time of the colonization of continents by life. Fossils of insects formed 100 million years later suggested that the present oxygen was practically attained at that time. Calculating in the percentage of the actual photosynthetic oxygen flux, in the Archean (between 4.5 and 2.5 billion years ago) the sources released 10% of oxygen; 9% were incorporated into rocks, and only 1% was used by biologic consumers. Presently, this latter figure, in the same units, is equal to 99.9% (Lovelock, 1988), but as a whole, only a small amount of photosynthetic oxygen is stored in the atmosphere-ocean system (5%). The major part can be found in iron oxides and sulfates (Schidlowski, 1978).

It should be noted that while photosynthesis makes free oxygen, decay consumes it (see Reaction 1.1). Net oxygen is produced only if fixed carbon is buried in sediments. Thus, the sediment formation, which was also the result of the presence of liquid water, has always been in close relationship with the oxygen concentration. On the other hand, due to sediment formation, carbon is stored on the Earth in the form of carbonates and not as CO_2 in the atmosphere as in the case of Venus.

At present atmospheric oxygen sources and sinks are in equilibrium (see Figure 1.1). Photosynthetic O_2 production is equal to the amount removed from the air by decay and respiration. The residence time (the ratio of the mass in the reservoir to its formation or removal rate) of oxygen in the atmosphere is equal to 3.8×10^3 years. Figure 1.1 shows that weathering of sedimentary rocks is balanced by the sediment formation. While weathering is independent of atmospheric oxygen pressure, sediment formation depends on O_2 concentration; the higher the oxygen level, the smaller the

---

* Note that the critical oxygen level and the time of the development of an ozone shield is a bit questionable due to the lack of our knowledge of the concentration of gases (e.g., N_2O) which remove O_3 molecules (see Section 3.2).

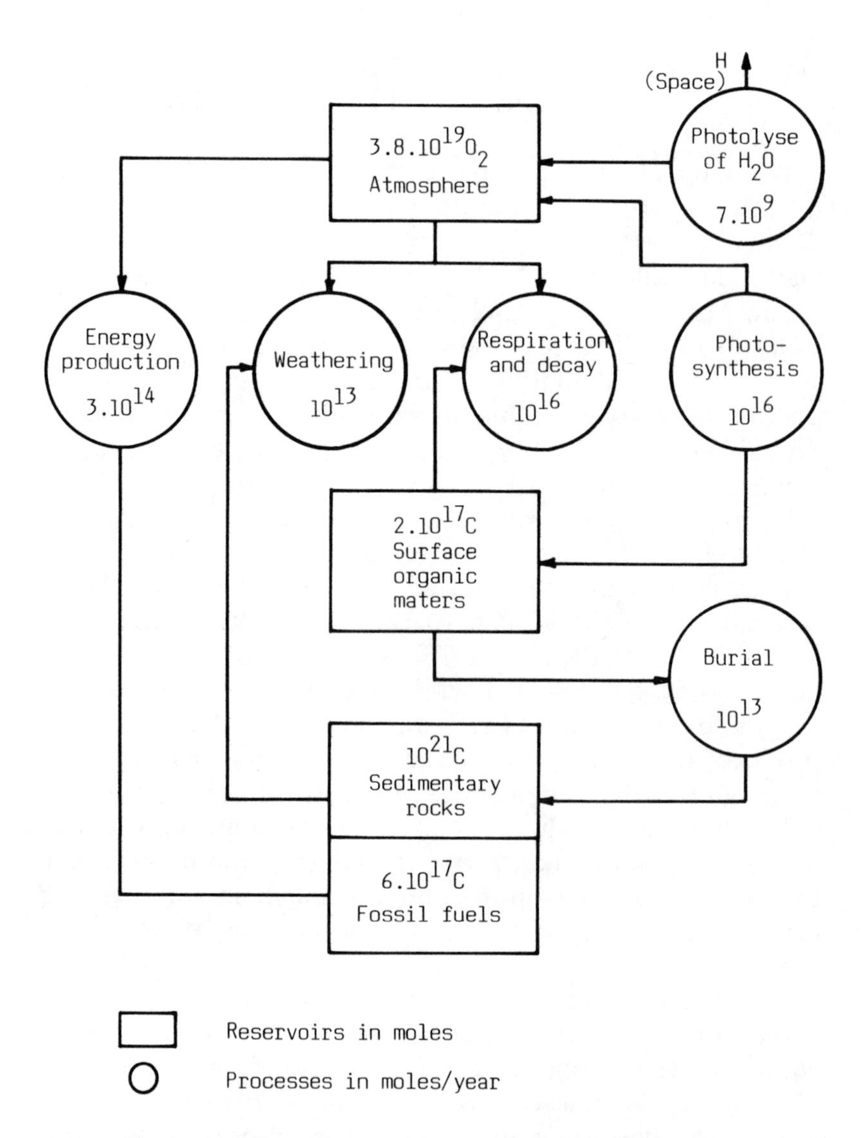

Figure 1.1 The cycle of oxygen in nature (Walker, 1977).

surface of anoxic areas. Consequently less organic carbon is buried in sediments, which leads to the decrease of O_2 pressure (Walker, 1974). The magnitude of fossil fuels used yearly by mankind seems to be rather important; however, it is very low relative to the oxygen amount in the air. Even the total quantity of carbon stored in fossil fuels can be neglected compared to the atmospheric oxygen burden. One can conclude that man cannot modify the concentration of atmospheric O_2. This conclusion is true, in particular, if the negative feedback mentioned is considered. The amount of surface organic carbon is also small relative to atmospheric O_2 mass. This means that even if the photosynthesis were stopped, respiration would halt after the organic carbon reservoir was exhausted (~20 years) and not after consuming the oxygen.

It was mentioned in the previous subsection that N_2 of volcanic origin* accumulated in the anoxic atmosphere because of the chemical stability of this gas. However, in such an oxidizing medium as the atmosphere at present, N_2 is thermodynamically unstable. The stable form would be nitrogen in nitrate ions dissolved in ocean water. It can be calculated (Warner, 1988) that the mass of N_2 would be reduced by 11% by this process. Lovelock (1988) speculates that this reaction is hindered by the biosphere — more exactly by its self-regulating capacity.

1.3 COMPOSITION-CLIMATE RELATIONSHIP DURING GEOLOGICAL TIMES

When life began the average temperature had to be very similar to the present value. This was a necessary condition, since life formation required a temperature between 0 and 50°C. However, considering the evolution of stars, it is probable that at that time the radiation coming from the sun was 25% less than it is now. We can assume that the temperature was held in the proper range by the presence of greenhouse gases like CO_2. Thus, Walker et al. (1981) propose that under abiologic conditions, the strength of the carbon dioxide sink, the sediment formation, depends solely on the temperature. If the temperature is lower, the weathering of silicate that contains rock necessary for carbonate formation is less effective, which leads to the accumulation of CO_2 in the atmosphere (negative feedback). Owen et al. (1979) calculated the CO_2 pressures necessary for proper temperatures by means of a simple climate model. According to their results, the necessary carbon dioxide concentration 4.25 billion years ago was 1000 times more than the actual level. A temperature of

* Lovelock (1988) proposed that the biosphere also produced nitrogen after life began.

20°C required a 200 times higher CO_2 pressure 3.5 billion years ago than it is now.

When life formed certain bacteria (methanogenics) released methane into the atmosphere. For this reason CH_4 concentration in the Archean was relatively high (Lovelock, 1988). Since CH_4 is also a greenhouse gas, this further promoted the increase of the temperature. However, when free O_2 appeared in the atmosphere 2.5 million years ago, methane concentration was decreased by oxidation.

The chemical weathering depends on the area of continental rocks exposed to the atmosphere (Holland, 1984). In other words, if the sea level is high and the continents are flooded, the intensity of weathering is low. In such periods the pressure of gases like CO_2 begins to rise, which increases the temperature. Figure 1.2 represents the changes in temperature and precipitation during geological times, reproduced on the basis of different evidence like the extension of continents (Frakes, 1979). Warm periods normally occurred in the case of marine transgression, while smaller sea surface resulted in colder climate. Thus, the temperature decrease after the end of Cretaceous Era coincides well with the emergence of the continents. After the formation of land plants (Carboniferous Era) the biosphere also became a regulator of chemical weathering. Generally speaking, dense vegetation promoted the chemical weathering, because it increased the CO_2 content of soil by respiration and decay of organic matters. By means of a suitable model, Berner (1990) calculated the atmospheric carbon concentration for the past 570 million years (Figure 1.3). Besides Reaction 1.1 and organic burial, he considered the following two processes:

$$CO_2 + CaSiO_3 \underset{\substack{\text{metamorphism} \\ \text{magmatism}}}{\overset{\text{weathering}}{\rightleftarrows}} CaCO_3 + SiO_2 \tag{1.2}$$

$$CO_2 + MgSiO_3 \underset{\substack{\text{metamorphism} \\ \text{magmatism}}}{\overset{\text{weathering}}{\rightleftarrows}} MgCO_3 + SiO_2 \tag{1.3}$$

This means that the weathering of silicate and carbonate rocks, as well as the thermal decomposition of minerals in great depths (magmatism, metamorphism), were taken into account.

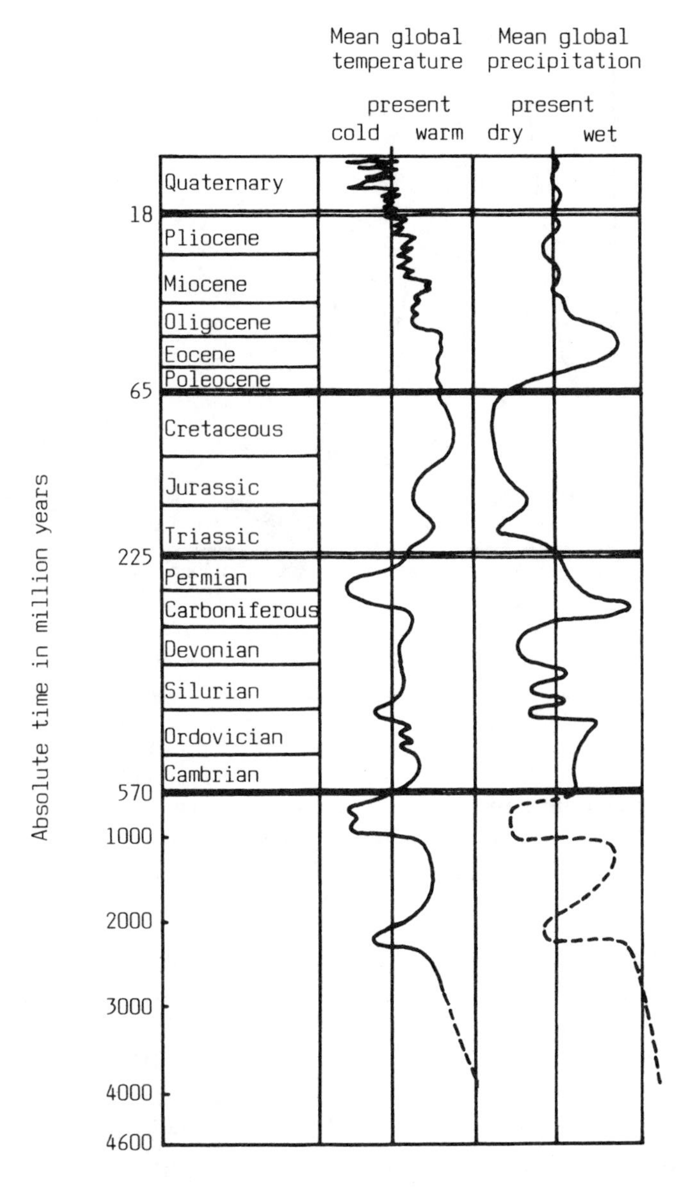

Figure 1.2 Generalized temperature history of the Earth (Frakes, 1979). Note that the time scale is not linear.

By comparing Figures 1.2 and 1.3 one can conclude that warm periods occurred when the carbon dioxide level in the atmosphere was high.

This conclusion is supported by the analyses of Antarctic ice samples aiming to estimate changes in atmospheric chemical

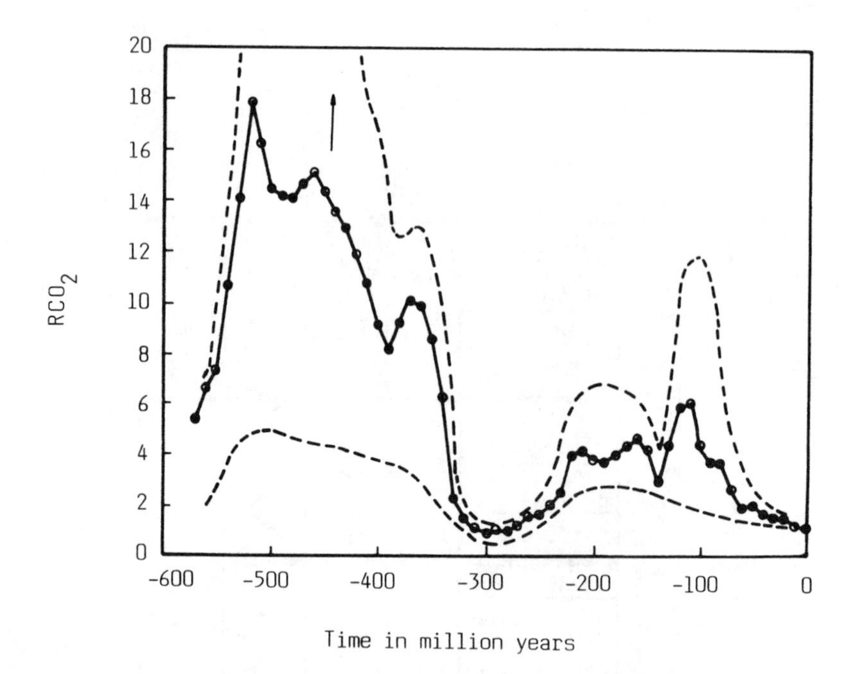

Figure 1.3 Standard plot vs time of RCO_2, the ratio of mass of CO_2 in the atmosphere at time t to that in the present atmosphere based on the carbon mass balance model of the present study. The solid curve represents the best estimate of the various parameters that go into the model. Dashed lines show the envelop of approximate error based on sensitivity analysis. The vertical arrow denotes that early Paleozoic CO_2 levels may have been even higher than those shown; see text for discussion (Berner, 1990).

composition and temperature during the last periods of the Earth's history. In these studies the temperature is determined on the basis of the ratio of oxygen isotopes in ice taken from different depths (time periods in the past). On the other hand, the composition of the atmosphere is estimated by means of the chemical analyses of air bubbles imbedded in ice samples. Figure 1.4 summarizes the results of such investigations (Chapellaz et al., 1990). It can be seen that the concentration of CO_2 and CH_4 varied parallel to the temperature during the last 160,000 years. Thus, in the last ice age (about 20,000 years ago) carbon dioxide and methane levels in the air were 190 ppm (10^{-6} part in volume) and 300 ppb (10^{-9} part in volume), respectively. In the Holocene (last 15,000 years) CH_4 concentration at least doubled, while CO_2 level reached a value around 280 ppm. These changes in concentration of greenhouse gases

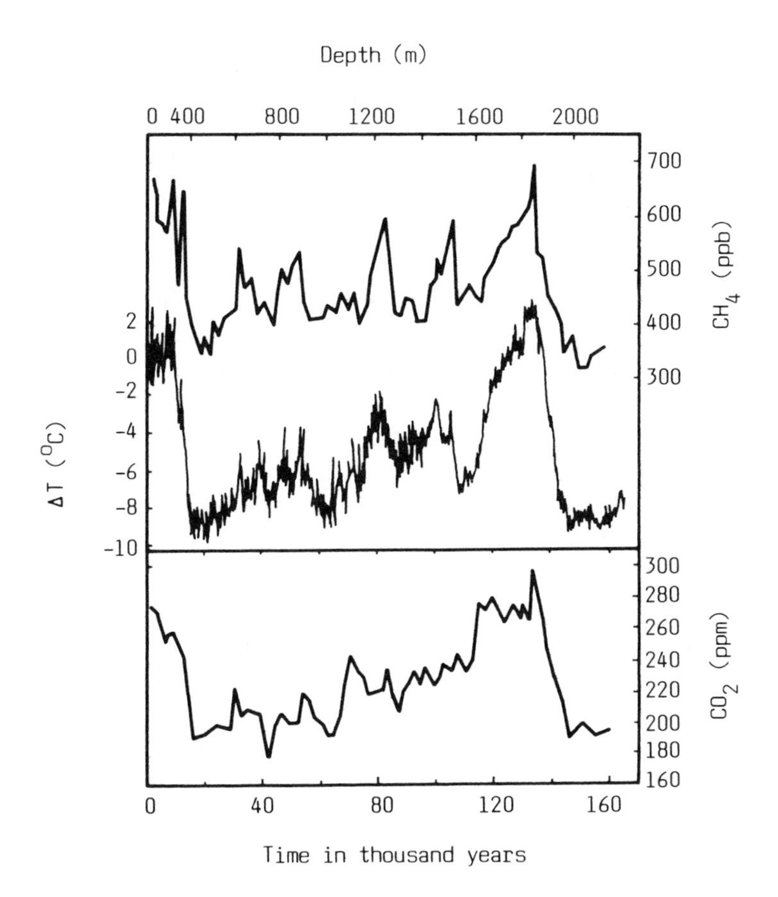

Figure 1.4 Results of analysis of Antarctic ice core records. Upper curve: methane concentration; curve at middle: temperature (ΔT) relative to present; lower curve: carbon dioxide level (Chapellaz et al., 1990).

coincided with an important warming of the climate of our planet. Model calculations suggest that the rise of atmospheric CO_2 alone was sufficient to explain this warming and the present day ice conditions (Lindstrom and MacAyeal, 1989).

Studies made in this field also demonstrated that about 14,000 years ago the concentration of N_2O was equal to 244 ppb, and 4000 years ago its value was 271 ppb (Zardini et al., 1989). According to another investigation (Khalil and Rasmussen, 1989) 3000 years ago the nitrous oxide level in the air varied around 285 ppb.

The atmosphere contains not only gases but solid and liquid aerosol particles as well. As it was mentioned, these particles scatter and absorb solar radiation. Consequently, higher aerosol content results in cooler temperature near the surface. On the basis that an important part of the particles is composed of sulfates (see also Section 3.4), it was proposed (Shaw, 1987) that the variation in sulfur cycle have participated in the regulation of the climate system. Since before the industrial era non-sea salt sulfates in the air were formed solely from gases of biological origin; one can speculate that the biosphere has regulated the Earth's temperature through the emission of sulfur gases. This idea was further developed by Charlson et al. (1987) arguing that the emission of biogenic gases controls not only the aerosol burden but also the structure of clouds, which determine, together with the extension of cloud cover, the albedo of the planet. The albedo is higher if the same water vapor quantity condenses on more particles/nuclei, resulting in smaller droplets of higher concentration (Twomey, 1977). The aerosol-climate feedback is negative if during warmer periods the oceanic biosphere emits more sulfur gases. However, the analyses of Antarctic ice cores indicate an increase in sulfate concentration under glacial conditions relative to interglacial periods (Legrand et al., 1988). This makes it possible for sulfate particles to have contributed to the formation of ice ages.

Finally, it should be noted that there is no intention here to suggest that the chemical composition of the atmosphere is the only factor which controls climate. Thus, the survey of the studies of oceanic sediments formed during the last 800,000 years shows that the temperature (and the composition, e.g., CO_2 concentration) has varied in periods that can be related to the fluctuation of the Earth's orbit around the sun (McElroy, 1986). On this basis one can speculate that the primary cause of climate changes has been the variation of the intensity of solar radiation owing to these fluctuations. However, climate changes in the past cannot be explained by the orbit fluctuations alone. The evidence presented above shows that atmospheric composition has always played an important role in the control of climate conditions. This means that present modification of the composition of the air by human activities constitutes a real threat for our future climate.

2

Present Variations of the Atmospheric Concentrations and Emissions of Trace Substances

2.1 THE STATEMENT OF THE PROBLEM: CHANGES IN CONCENTRATION

Recently, changes in atmospheric composition have been observed in time periods much shorter than the geological scale discussed in Chapter 1. During the last 200 years the concentration of several trace constituents in the air has changed at a rate unknown in the past. *Trace gases* have a very low concentration; they constitute <0.04% of the volume of the dry atmosphere.* However, they play an essential role in the control of climate and atmospheric chemical processes. Due to their low levels, human activities can modify their concentration by releasing pollutants into the air. This causes perturbations in the environment.

The effects of mankind are demonstrated partly by ice sample analyses and partly by direct atmospheric observations carried out during the last decades. In Figure 2.1 the results of carbon

* The composition is dominated by the main constituents: nitrogen (78.080%), oxygen (20.948%), and argon (0.934%).

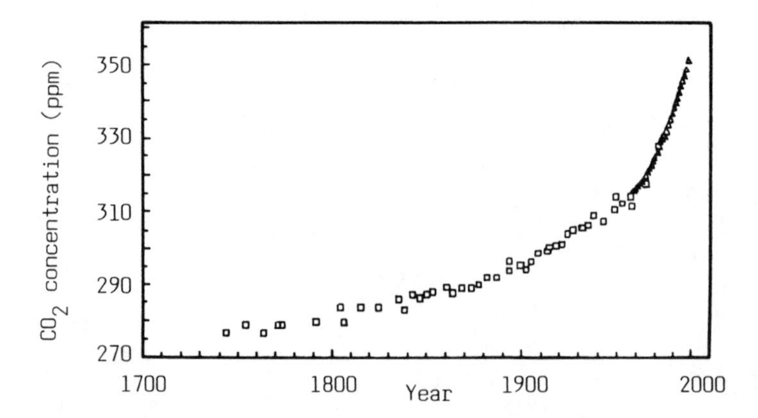

Figure 2.1a Atmospheric CO_2 increase in the past 250 years, as indicated by measurements on air trapped in ice from Siple Station, Antarctica (data compiled in IPCC, 1990).

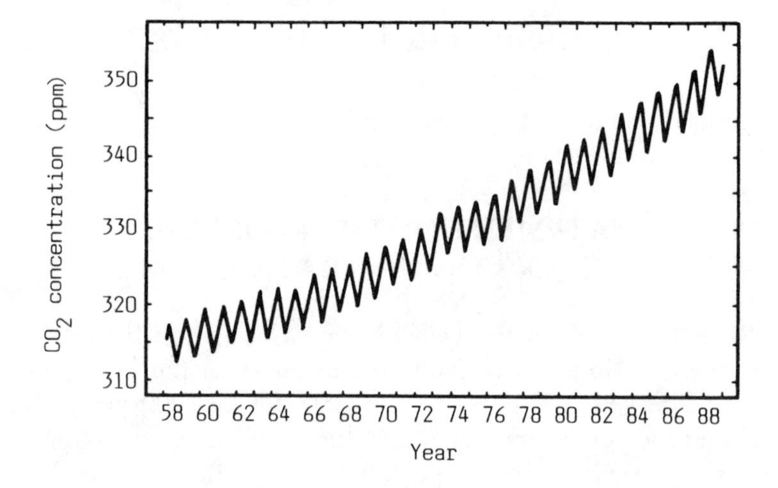

Figure 2.1b Monthly average CO_2 concentration in parts per million of dry air, observed continuously at Mauna Loa, Hawaii (Keeling et al., 1989). The seasonal variations are due primarily to the withdrawal and production of CO_2 by the terrestrial biota.

dioxide measurements are combined. Except for its role in the regulation of the acid-base equilibrium of cloud and precipitation waters, this gas is chemically inert. However, CO_2 is a biologically active compound as indicated by the seasonal variation of its concentration, showing a minimum during the vegetation period. Further, being a greenhouse gas, CO_2 absorbs the longwave radiation that influences the Earth's

Table 2.1 The Most Important Greenhouse Gases Influenced by Human Activities[a]

Parameter	CO_2 (ppm)	CH_4 (ppm)	CFC-11 (ppb)	CFC-12 (ppb)	N_2O (ppb)
Preindustrial concentration	280	0.8	0	0	288
Current concentration	353	1.72	0.280	0.484	310
Current rate of increase	1.8 (0.5%)	0.015 (0.9%)	0.0095 (4%)	0.017 (4%)	0.8 (0.25%)
Residence time in years	(50–200)	10	65	130	150

Note: For the explanation of the range of CO_2 residence time see Section 5.2.1.

[a] Intergovernmental Panel on Climate Change (IPCC) 1990.

radiation balance. One can see from Figure 2.1 that while the concentration of CO_2 in the air was about 280 ppm before the industrial revolution, its present level is around 350 ppm. Thus, during 200 years the concentration has increased by 70 ppm, which is equivalent to an average positive rate of 0.35 ppm yr^{-1}. The current rate of increase reaches a value of 1.8 ppm yr^{-1} (0.5% yr^{-1}).

As Table 2.1 shows, the concentration of other biogenic greenhouse gases like methane and nitrous oxide has increased since the beginning of the industrial revolution.* This can modify not only the absorption of longwave radiation but also the ozone formation in the stratosphere (by N_2O) and troposphere (by CH_4). Data indicate that even such species like CFCs have been emitted into the atmosphere; they did not exist there before human modifications.** Such a class of substances is constituted by different chlorofluorocarbons (CFCs, also called freons). Not only are greenhouse gases, but they pose a threat to the stratospheric ozone layer. The residence times, also presented in Table 2.1, are rather long. This means

* Water vapor in the troposphere is the most important greenhouse gas, however, its concentration is not influenced directly by human activities (see also Section 6.5).

** Note that recent Russian observations suggest that some CFCs are released by volcanoes (Isidorov et al., 1990).

that the concentration of these gases follows changes in emissions relatively slowly.

Under undisturbed conditions an important part of sulfur and nitrogen containing gases (other than N_2O) are also released into the air by the biosphere. A certain part of natural sulfur gas emission is caused by volcanic eruptions, while the balance of nitrogen oxide emissions is due to the combination of N_2 and O_2 during high temperature (e.g., lightning discharges) and to biomass burning. Sulfur and nitrogen gases are reactive species; they are converted in the air into acid vapors and aerosols. Consequently, these species play an essential role in the control of the acidity of atmospheric depositions. The study of the composition in snow from Greenland demonstrates that the nitrate and sulfate concentration in deposition has increased since the end of the last century (Nefter et al., 1985), while it has remained constant (except for natural variations) over Antarctica (Legrand and Delmas, 1984). This difference between the two hemispheres is caused partly by the distribution of anthropogenic sources and partly by the short atmospheric residence time of sulfur and nitrogen compounds.

The chemical transformation of organics, as well as sulfur and nitrogen containing gases, is caused by the presence of free radicals formed mainly from ozone and water vapor under the influence of solar radiation. The comparison of ozone measurements of the surface air carried out during the last century and the present day makes it evident that human activities have increased the concentration of tropospheric ozone (Volz and Kley, 1988), which has also contributed to the man-made greenhouse effect.

This short discussion shows that the composition of the atmosphere has changed considerably during the last two centuries. The reasons are human activities like energy production, industry, transport, and agriculture. It can be demonstrated that the present concentration variation of greenhouse gases is proportional to the increase of world populations (Crutzen and Brühl, 1989). In order to understand the causes of these variations in composition, the strengths of different natural and anthropogenic sources should be reviewed and compared. This procedure makes it possible to gain further insight into the problem of atmospheric environment modification caused by mankind.

2.2 BIOGENIC EMISSIONS OF TRACE MATERIALS

2.2.1 General

The aim of this section is to evaluate the material release into the air from biospheric sources. It goes without saying that on the continents man has modified this release with agricultural practice. The measure of the modification varies from one source type to the other. Some emissions (e.g., from nitrification and denitrification) are only moderately affected. Others, like methane and ammonia production, are substantially altered. Even material fluxes caused by deforestation and biomass burning are caused by human activities.

2.2.2 Oceanic Emissions of Trace Gases

As it was discussed in Chapter 1, different media of our environment have developed in close interaction. The relationship between the evolution of the atmosphere and oceans has been interrelated in particular; even in our time the level of many atmospheric constituents is controlled by the release of gases from the oceans. Besides oxygen, the oceans emit several gases containing sulfur, halogens, carbon, and nitrogen, which can be found as trace compounds in the air. They influence significantly chemical processes in the atmosphere as well as the climate of the planet.

The most important sulfur gases emitted by the oceans are *dimethyl sulfide* (DMS) and *carbonyl sulfide* (COS), which are converted into aerosol particles (see Section 3.4) in the troposphere and stratosphere, respectively. In this way the emission of these gases is related to climate regulation and cloud formation as will be discussed in Section 6.4.

DMS is produced by algae in the upper 50 m of the oceans, which receive sufficient amounts of solar radiation for photosynthetic growth processes. It is speculated that DMS forms mostly by the bacterial decomposition of dimethyl propiothetin released by aged cells (Warneck, 1988). In spite of the fact that the oceanic concentration of DMS is quite low, the water is supersaturated with respect to the atmosphere. Moreover, DMS can be detected everywhere in the surface water which explains its relatively high emission. Andreae (1986) estimates that this emission is equal to 32 Tg S yr^{-1} (1 Tg = 10^{12} g). On the other hand, Bates et al. (1987) propose half this quantity. The discrepancy is caused by the different transfer rates between ocean and atmosphere assumed by the authors mentioned. However, considering the distribution of the world ocean between the two hemispheres it is evident that the larger part of DMS is liberated into the air in the Southern Hemisphere.

COS is not released directly by the life process of planktonic algae. It is formed by the photochemical reaction of oxygen and sulfur containing organics. Due to the photochemical production of COS, its emission is higher during the day than during the night. Rasmussen et al. (1982) measured the concentration of COS in the ocean water and the surface air and deduced an emission rate of 0.6 ± 0.3 Tg yr^{-1}. This means that the oceans provide a small but net source of atmospheric COS. In contrast to DMS, COS is an inert compound in the troposphere. Consequently, it reaches the stratosphere, where its molecules are destroyed photochemically under the effect of ultraviolet radiation.

Among halogen compounds emitted by the oceans, *methyl chloride* should be mentioned first; it plays an important role in the control of the stratospheric ozone layer (see Section 3.2). Its emission by seaweeds has been demonstrated by several studies. It is probable, however, that CH_3Cl molecules are also produced in the upper layer of the oceans, e.g., by the interaction of dimethyl propiothetin and iodide ions in the sea water. One can estimate that 3 Tg methyl chloride is emitted yearly to the atmosphere from the entire ocean surface (Singh et al., 1979). In lesser amount, methyl iodide and methyl bromide are also released by oceanic sources.

Besides COS and CH_3Cl, the oceans also liberate other gaseous species. Among the inorganic gases emitted, *carbon monoxide* is the dominant compound. While the origin of oceanic CO molecules is not entirely clear, we can assume that CO arises from the photooxidation of organic compounds released by algae (see Warneck, 1988). The range of the possible emission rate is centered around 100 Tg yr^{-1} (43 Tg C yr^{-1}), proposed originally by Seiler (1974).

As measurements made in the air and the water show, *methane* is also emitted by the ocean in a quantity between 3 Tg C yr^{-1} (Sheppard et al., 1982) and 7.5 Tg C yr^{-1} (Cicerone and Oremland, 1988). Near the shore this emission is obviously due to the bacterial degradation of organic matter. The reason for the supersaturation of open ocean water with respect to the atmosphere is not well understood. It is known, however, that ocean sediments contain large amounts of methane, and it cannot be excluded that a part of this methane quantity is liberated into the water and subsequently to the atmosphere. Although the amount of oceanic release is low compared to CH_4 emission of continental sources (see Section 2.2.4), methane molecules of oceanic origin contribute to the control of the atmospheric level of this important greenhouse gas.

Observations carried out in oceanic environment indicate that another greenhouse gas, *nitrous oxide,* is also produced in the ocean. Cohen and Gordon (1979) speculate that oceanic N_2O production is due to the nitrification processes discussed in the next subsection. These authors propose that the strength of the oceanic N_2O source is between 4 and 10 Tg N yr^{-1}, while more recent studies (IPCC, 1990) makes possible a lower emission range (1.4 to 2.6 Tg N yr^{-1}).

Finally, it should be noted that *carbon dioxide* is also liberated from the oceans to the atmosphere in a large quantity. However, the oceans do not provide a net CO_2 source, since the biological carbon cycle over the ocean is closed. Rather, the oceans provide a CO_2 sink as discussed in Section 5.2.

2.2.3 Nitrification and Denitrification in Terrestrial Ecosystems

Nitrification and denitrification processes play an important role in the regulation of the cycle of nitrogen compounds in nature. Nitrification is defined as the biological conversion of nitrogen species from a reduced to a more oxidized state, like the oxidation of ammonium to nitrite and nitrate. On the other hand, denitrification denotes the inverse reaction, that is the conversion of nitrate into ammonium ions, gaseous nitrogen, or nitrous oxides. It goes without saying that nitrification and denitrification in terrestrial ecosystems are modified by man — by land use practice and the use of fertilizers.

Nitrous oxide is the most abundant nitrogen compound in the atmosphere. It is produced both by denitrification and nitrification. The most direct way of N_2O formation in soils is the reduction of nitrate ions by hydrogen in gaseous and ionic form (Delwiche, 1978). The process is significant in particular if the amount of organic substrate is limited but the amount of electron-accepting species is high. However, nitrifying microorganisms can also produce nitrous oxide during the nitrification of ammonium or hydroxylamine that comes from the decomposition of organic materials (Bremner and Blackmer, 1981). The emission of N_2O from different soils was found to be higher during the summer months (Bremner et al., 1980), and the release from soils treated with nitrogen fertilizers in the form of ammonium sulfate or urea was significant in particular (Breitenbeck et al., 1980). However, the increase in emission due to fertilizers occurs only for a limited time period during the year. According to recent estimates (IPCC, 1990) the global soil N_2O–N source strength varies between 2.9 Tg yr^{-1} and 5.2 Tg yr^{-1}. The range is caused by the uncertainties of such a calculation. Estimating the effect of fertilizer use on the N_2O emission is even more complicated. The values published are between 1 and 40% of the above figures.

Laboratory experiments demonstrate that nitrification also produces *nitric oxide* (Lipschultz et al., 1981) under appropriate conditions. It was demonstrated (see Penner et al., 1991) that denitrification also produces NO in the anaerobic environment. According to the estimate of these latter authors soil microbial activity emits about 10 Tg of nitrogen annually in the form of nitric oxide; 79% of this amount is released by tropical forests. One can speculate that only a small fraction of this emission is due to the use of fertilizers.

2.2.4 Decay of Organic Materials

As discussed in Section 1.2.2, the decay of organic substances formed by photosynthesis produces carbon dioxide if oxygen is available. This means that the carbon used by plants is returned to the atmosphere as CO_2. More exactly, leaves and other assimilating parts of plants are decayed in the litter by using oxygen molecules. This liberates carbon dioxide into the air either directly or indirectly through the soil humus. Figure 2.2 represents fluxes of organic carbon in the terrestrial biosphere by means of a four reservoir model (Warneck, 1988). It can be seen that the total carbon quantity released directly from the litter by microbial decay is 42 Pg yr^{-1} (1 Pg = 10^{15} g), while humus emits annually 10 Pg C into the atmosphere. The sum of these two terms is nearly equal to the CO_2 mass re-

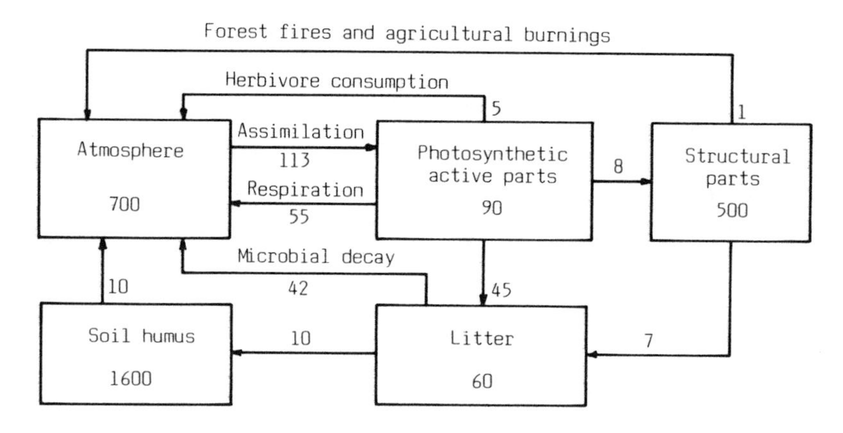

Figure 2.2 Four reservoir model of organic carbon in the terrestrial biosphere. Reservoir contents are expressed in units of Pg carbon ($=10^{12}$ kg C) and fluxes in units of Pg carbon per year (Warneck, 1988).

leased by respiration, while the consumption of plants by herbivorous animals produces about 10% of this quantity. Forest fires and agricultural burning by man (see Section 2.2.6) provide a relatively low source strength compared to decay and respiration.*

If the bacterial degradation takes place under anaerob (oxygen deficient) conditions methane is generated and liberated into the atmosphere, where it acts as a greenhouse gas before being destroyed chemically. Anaerob conditions can be found in sludges or sediments of lakes and swamps and in the rumen of cattle. It was proposed that CH_4 molecules are even created in the digestive tract of herbivorous insects like termites.

Human activities can modify biogenic methane emissions considerably by increasing the number of domestic animals and extending the area of rice paddy fields. Several estimates can be found in the literature concerning the strength of different terrestrial biogenic methane sources (see e.g., Warneck, 1988). Estimates agree within a factor of 2 to 3. The most acceptable values for rice paddies and natural wetlands (including tundra) are 83 Tg C yr^{-1} and 86 Tg C yr^{-1}, respectively (Cicerone and Oremland, 1988). The corresponding figure for enteric fermentation of animals lies between 49 and 75 Tg C yr^{-1} with an

* Due to recent calculations by Crutzen and Andreae (1990) this statement is questionable; more studies have to be done in this field.

Table 2.2 Terrestrial Biogenic Sources of Atmospheric Methane[a]

Sources	Annual Release[b]	Range[b]
Natural wetlands	86	75–150
Rice paddies	83	19–128
Enteric fermentation	60	49–75
Termites	30	7–75
Σ	259	150–428

[a] Cicerone and Oremland, 1988; IPCC, 1990.
[b] The values are expressed in Tg of carbon per year.

average of 60 Tg C yr^{-1}. The methane quantity emitted by termites is uncertain in particular. Values between 1.5 and 112 Tg C yr^{-1} have been published (Warneck, 1988). The most probable range is 7.5 to 75 Tg C yr^{-1} with a mean value of 30 Tg C yr^{-1}. Table 2.2 summarizes the above discussion. One can see that the total emission from terrestrial ecosystems is 259 Tg C yr^{-1}, with an anthropogenic contribution of about 50%.

When nitrogen containing organic materials are decayed *ammonia gas* is produced, which controls in a significant way the acidity of the air and precipitation. The decay can take place in the soil and on the surface where the decomposition of animal urine releases a large quantity of NH_3 molecules. On the basis of the survey of different publications Warneck (1988) estimates that the soil emission is 15 Tg N yr^{-1}, while domestic and wild animals produce 22 and 4 Tg yr^{-1} of NH_3–N on a global scale. The ammonia release from human excrements is estimated to be 3 Tg N yr^{-1}. Field studies also show that 5 to 10% of nitrogen of fertilizers is lost to the air as NH_3. Combining this figure with the world fertilizer production, the calculation results in a global ammonia emission of 3 Tg yr^{-1}, which increases the soil emission with about 20%.

Finally, we note that among the end products of the degradation of organic matter *hydrogen sulfide, carbon disulfide,* and *organic sulfides* (DMS, COS) can also be found since sulfur is an essential element of living organisms. Sulfur gas emission from soils is relatively high in coastal marshlands and under tropical conditions. Thus, reduced sulfur emission of lands between 25° N and 25° S is between 5 and 10 Tg S yr^{-1}, while continental areas outside this region emit 3 Tg S yr^{-1} (Warneck,

1988). It is recently speculated, however, that this source term is lower. Thus, Langner and Rodhe (1991) accepted a value of 1 Tg S yr^{-1} for their model of the tropospheric sulfur cycle. They assume that this sulfur amount is mostly due to DMS molecules. Further research is needed to determine the reason for this discrepancy.

2.2.5 Direct Emissions from the Vegetation

Aside from respiration releasing carbon dioxide (see above), the vegetation emits directly into the air other carbon containing gases like carbon monoxide and different organics, mostly nonmethane hydrocarbons (NMHC).

Experiments show that green plants can both absorb and liberate carbon monoxide. However, in the case of each plant studied, a net production of carbon monoxide was observed depending on light intensity. According to Seiler and Conrad (1987) the global strength of this source is around 75 Tg yr^{-1} (32 Tg yr^{-1} in carbon equivalents).

A much higher quantity of volatile carbon is emitted by the vegetation, particularly by deciduous trees and conifers. While deciduous trees (e.g., oak) emit isoprene, conifers release terpenes. Isoprene emission is light dependent, but the terpene flux is independent of radiation intensity. According to Zimmerman et al. (1978) NMHC emission is 0.7% of net primary productivity. By using this ratio a global hydrocarbon emission of 830 Tg yr^{-1} can be obtained. More recently Fehsenfeld (1990) calculated the global organic carbon release by the vegetation and obtained a figure of 700 Tg yr^{-1}. Fehsenfeld proposes that 80% of this emission is isoprene, which is twice corresponding ratio published by Zimmerman et al. (1978). Fehsenfeld concluded that the biogenic hydrocarbon production is higher than the anthropogenic emission (see later) mainly during summertime. This means that organic species of natural origin control significantly the chemistry of hydrocarbons in the atmosphere.

The carbon emission of the biosphere in *aerosol* form was estimated by Cachier et al. (1986). On the basis of their studies they propose a global source strength of 12 Tg C yr^{-1}. However, it is not clear whether this amount is directly emitted by the vegetation (primary particles) or formed in the air from organic vapors of biospheric origin by gas-to-particle conversion (secondary particles).

The biosphere also releases different *metals* to the atmosphere. Although the quantity of these emissions is not too large in absolute terms, 30 to 50% of the total natural emissions of cadmium, lead, copper, and zinc are due to biogenic sources as discussed by Nriagu (1989) (see also Section 2.5).

2.2.6 Deforestation and Biomass Burning

One of the most determining factors of human influences on natural vegetation is the clearing of forests to transform the surface to cultivated or pastoral lands. This activity makes an important change in the carbon content of the biosphere, since trees and soils in forests contain 20 to 100 times more carbon per unit area than agricultural systems. The flux of carbon in the form of CO_2 into the air caused by deforestation is due to burning, decay of biomass on the site (e.g., roots, twigs), oxidation of wood products removed (paper, lumber etc.), and oxidation of soil carbon (IPCC, 1990).

While in the past deforestation on mid-latitudes was the controlling process, the clearing of tropical forests is presently the main area of this activity. Detwiler and Hall (1988) give global average CO_2–C release of 1 Pg yr^{-1} caused by the deforestation in tropical regions. On the other hand they suggest that in 1980 the effect of the clearing of nontropical forests on CO_2 flux was practically nonexistent. It should be noted, however, that the value proposed for the clearing of tropical forests is not well established. Thus, Houghton et al. (1987) propose for this term a value of 1.8 Pg C yr^{-1}, which is nearly double the figure of Detwiler and Hall. Anyway, until 1980 at least 90 Pg C was released into the air by human changes in vegetation cover (Siegenthaler and Oeschger, 1987).

Biomass burning produces not only carbon dioxide, but also carbon monoxide, methane, and many other atmospheric trace gases and aerosol particles. Crutzen and Andreae (1990) estimate that biomass containing 2 to 5 Pg C is burned annually by man, mostly in developing countries. While the major part of carbon is released into the air as CO_2, 10 ± 5% of carbon content is liberated as CO and 1 ± 0.6% as methane. Table 2.3 gives the emission of different gases and aerosol particles due to biomass burning. In the table, the data of Crutzen and

Table 2.3 Emission of Different Trace Materials from Biomass Burning[a]

Species[b]	Emission[c]
CO	120–510
CH_4	11–53
NMHC	34
CH_3Cl	0.5–2.0
NO_x (x = 1, 2)	2.1–5.5
NH_3	0.5–2.0
N_2O	0.1–0.3
SO_2	1.0–4.0
COS	<0.1–0.2
TPM	36–154
POC	24–102
EC	6.4–28
K	0.5–21

[a] Data taken from Crutzen and Andreae (1990) except that for nonmethane hydrocarbons (Greenberg et al., 1984).
[b] NMHC: nonmethane hydrocarbons, TPM: total particulate matter, POC: particulate organic carbon, EC: elemental carbon.
[c] Values expressed in Tg of C, CH3Cl, N, and S per year.

Andreae (1990) are completed by the emission rate of nonmethane hydrocarbons given by Greenberg et al. (1984) on the basis of their research in Brazil. It can be seen that biomass burning liberates gases and aerosol particles that influence the climate (greenhouse gases and aerosol particles), stratospheric (CH_3Cl) and tropospheric (NO_x, hydrocarbons) ozone level, as well as the acidity of the environment (SO_2, NO_x, NH_3). It is notable that during burning a relatively high potassium mass is emitted. Thus, we can conclude that biomass burning not only transforms the land surface, but it contributes significantly to changes in atmospheric composition.

2.3 A NONBIOLOGICAL NATURAL SOURCE: THE VOLCANIC ACTIVITY

As discussion in Section 1.2 shows, volcanic gases have played an important part in controlling the development of the early atmosphere. Although the role of these gases is not as important now as in the past, the effects of volcanic activity cannot be neglected. This is true in particular in the case of sulfur species, since sulfur gases of volcanic origin control in a large measure the stratospheric aerosol layer (see Section 3.4.3).

As we discussed in Section 1.2.1, volcanic gases are composed mainly of H_2O, CO_2, N_2, and SO_2.* The emission of the first three gases mentioned can be neglected compared to the evaporation of the oceans and emissions from biogenic sources. The amount of *sulfur dioxide* is also not too large; however, SO_2 molecules are injected in this way into the stratosphere, where they serve as precursors for sulfuric acid aerosol particles. Consequently the determination of the strength of volcanic sulfur sources is crucial for the study of atmospheric composition.

In spite of this importance the sulfur mass emitted annually by volcanoes is rather uncertain. Earlier estimates (e.g., Friend, 1973) are around 1 to 2 Tg S yr^{-1}, while more recent studies suggest a higher value. This is caused by the fact that earlier investigations consider only gases emitted during the eruptions of volcanoes (Berresheim and Jaeschke, 1983). However, the number of live volcanoes is about an order of magnitude higher than the number of eruptions per year (55 on an average). If posteruptive volcanoes are also taken into account, an emission rate of 7 Tg S yr^{-1} can be obtained (Berresheim and Jaeschke, 1983). Thus, we can conclude that the best present estimate for total volcanic SO_2 emission is around 8 Tg S yr^{-1}. It is possible, however, that the precision of this figure is not better than a factor of 2. Further, the authors referenced speculated that volcanoes also emit *sulfate* particles directly in the air. Their estimate for this flux is around 3 Tg S yr^{-1}.

Besides gases, *ash* particles are emitted during volcanic eruptions. Farlow et al. (1981) report that ash grains are composed mostly of different glasses. While fine sulfate particles formed by gas-to-particle conversion remain in the stratosphere for at least one year, large ash particles fall relatively quickly out of the atmosphere. However, they play an important temporary role in the control of atmospheric optical properties.

Finally, we note that some *metals* are also released by volcanoes. While the quantities emitted are not too significant in absolute terms, metals of volcanic origin can play a role, at least temporarily, in the control of atmospheric deposition of

* Note that some H_2S can also be found in volcanic gases (about 10% of total sulfur); or it is converted rapidly in the air to form SO_2..

microelements under clean conditions. As an example we mention that the amount of cadmium, copper, nickel, lead, vanadium, and zinc emitted annually by volcanoes is 0.11, 3.8, 2.3, 1.9, 1.6, and 7.6 Gg (1 Gg = 10^9 g) (Nriagu, 1989), respectively.

2.4 ENERGY PRODUCTION AND INDUSTRIAL ACTIVITIES

2.4.1 Energy Production

Energy production is a key issue of human life. At the present time about 75% of the total energy used by mankind is produced by the combustion of solid (carbon), liquid (oil), and gaseous (natural gas) fossil fuels. According to the data of the Worldwatch Institute (Washington, D.C.) in 1989 more than 7000 Mt in oil equivalents fossil fuels were burned. The share of carbon, coal, oil, and natural gas was 32, 44, and 24%, respectively.

While complete combustion produces carbon dioxide, during incomplete combustion carbon monoxide is released. Carbon emission to the atmosphere is a function of the type of fossil fuels. It is the lowest for natural gas (13.7 Tg C EJ^{-1}, where EJ designates exajoule = 10^{18} J) and highest for coal (23.8 Tg C EJ^{-1}). The corresponding figure for oil is 19.2 Tg C EJ^{-1} (MacCracken, 1990). At the same time the oxides of different elements contained in the fuels (like sulfur and nitrogen) are also emitted to the air.

Data compilation of the Oak Ridge National Laboratory shows that in 1860 the global CO_2–C emission was as low as 0.1 Pg (Boden et al., 1990). However, in 1988 the emission reached a value of 5.89 Pg. About 97% of this emission was due to fossil fuel burning, while the negligible balance was caused by cement manufacturing and gas flaring. During carbon and oil burning an equal amount of carbon was emitted (2.39 Pg yr^{-1}), while the use of natural gas yielded less emission (0.92 Tg yr^{-1}). Taking into account changes in world population one can calculate that from 1950 to 1988 the emission per capita increased from 0.7 to 1.2 t yr^{-1} on an average.

In Figure 2.3 the carbon dioxide emissions (expressed in carbon equivalent) of the different parts of the world are illustrated. It can be seen that before 1960 the more developed part of the world (North America, Western Europe) domi-

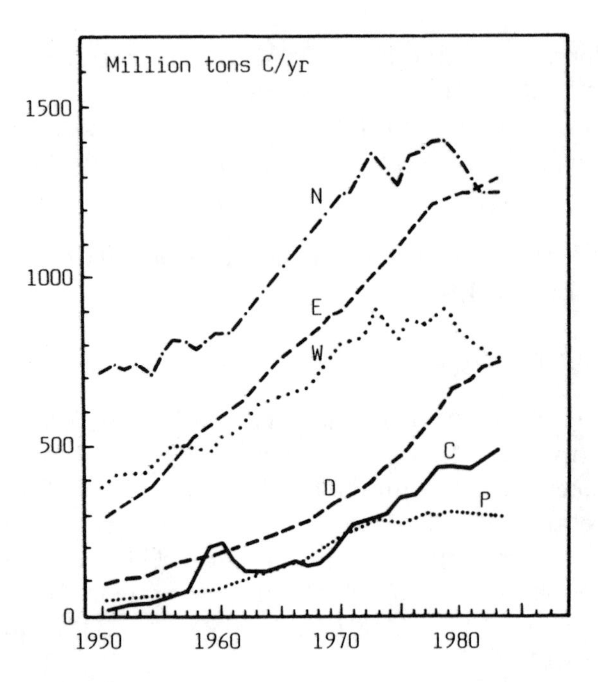

Figure 2.3 Variation in the annual CO_2 emissions in different regions of the world according to Rotty (1987). N: North America, E: Eastern Europe and Russia, W: Western Europe, D: developing countries, C: centrally planned Asia, P: Pacific.

nated the global emission. However, in North America and Western Europe the emission has been stabilized. At the same time the share of Eastern Europe and the developing world has increased considerably. Also, the emission in centrally planned Asia (mostly China) has exceeded the CO_2–C release over the Pacific (e.g., Japan). Considering the possible inadvertent climate modifications caused by the increase of atmospheric carbon dioxide levels further efforts have to be made to stabilize or reduce CO_2 emissions all over the world.

The carbon monoxide emitted annually is much less than the mass of CO_2 released by burning fossil fuels. However, since CO molecules are removed from the air mostly by their reaction with OH radicals, anthropogenic CO plays an important role in the control of tropospheric chemistry. Table 2.4 contains information published by Logan et al. (1981) for the years 1974 to 1978. One can see that, except for North America, the emission is dominated by the use of solid fossil fuel. The total emission is 125 Tg C yr^{-1}, which is between the CO amount emitted annually by the oceans (see Section 2.2.2)

Table 2.4 CO Emission from Fossil Fuel Burning in Different Parts of the World for 1974 and 1978[a]

Sources Type	Fuel Type	N. America[b]	Europe[b]	Rest of the World[b]	Total[b]
Stationary	Coal	0.4	10.0	10.0	20.4
	Lignite	—	1.3	—	1.3
	Gas	<0.1	0.1	<0.1	0.1
	Oil	0.9	2.0	1.6	4.5
	Subtotal	1.3	13.4	11.6	26.3
Mobile	Gasoline and oil	40	31	28	98
Total		41.3	44.4	39.6	125.3

[a] Logan et al., 1981.
[b] Values expressed in Tg of carbon per year.

and the amount released by biomass burning (Section 2.2.6). It should be noted, however, that such a comparison must be considered with caution due to the uncertainties involved in the estimation of different emission terms. Anyway, combustion of gasoline in motor vehicles provides an important global carbon monoxide source.

Different fossil fuels contain sulfur with a relative magnitude around 1%. During combustion an important fraction of this sulfur is liberated to the atmosphere as sulfur dioxide. The quantity released is a function of different parameters, like the type of boilers and the nature of fuel, as well as the emission control applied. Like CO_2 emission, the global sulfur dioxide emission has increased considerably during the last century. According to Dignon and Hameed (1989), in 1860 it was equal to 2.0 Tg S, while for 1980 a value of 62.8 Tg S was calculated by the same authors. The contribution of North America, Europe, and Asia to the total emission was 23, 24, and 45%, respectively, while the remaining part was emitted by the rest of the world. However, in 1930 the global emission of sulfur was dominated by the more industrialized North America and Europe.

During high temperature combustion processes nitric oxide is produced from the reaction of N_2 and O_2 in the air, as well as from oxidation of nitrogen in the fuel. The calculations of Dignon and Hameed (1989) show that in 1980 55 times more NO–N was emitted by burning fossil fuels than in 1860. For the emission in 1980 they give a total value of 22.1 Tg N, which is in agreement with the earlier estimate of Logan (1983). Logan also estimated that about 60% of this amount is emitted by stationary sources, while the balance is released during transportation. As for sulfur dioxide, the major part of nitrogen oxide is liberated to the atmosphere over the Northern Hemisphere, mostly in North America, Europe, and Asia. This means that in over-industrialized continents the atmospheric sulfur and nitrogen burdens are controlled entirely by anthropogenic processes.

Combustion of fossil fuels produces a large amount of different nonmethane hydrocarbons (NMHC). Like carbon monoxide the role of mobile sources is predominant, since in vehicular exhaust a large amount of NMHC is emitted to the air. According to the publication of Ehhalt et al. (1986), stationary combustion releases 4 Tg NMHC yr^{-1} globally. The corresponding figure for transportation is 20 Tg yr^{-1}.

It should be noted that gaseous species other than those discussed above are also emitted to the atmosphere during combustion. For example, nitrous oxide is mentioned, which is released globally in an amount of 0.1 to 0.3 Tg N yr^{-1} (IPCC, 1990). However, this amount can be practically neglected relative to other N_2O source terms. The same is true for ammonia, which is also produced during combustion in a small amount (Warneck, 1988).

Besides gaseous compounds, aerosol particles are also formed if different fossil fuels are burned. The most obvious example is the release of carbon containing (carbonaceous) particles, which can be composed either of organics or elemental carbon. Taking into account the size and isotopic composition of carbonaceous particles collected under different conditions, Cachier et al. (1986) concluded that 16 Tg C yr^{-1} particulate carbon is emitted to the atmosphere on a global scale. It is more than probable, however, that an important fraction of carbonaceous particles is not directly emitted but form in the atmosphere by chemical reactions and condensation as discussed in Section 3.4.2.

2.4.2 Industry and Mining

During industrial production a lot of substances in gaseous and aerosol form are released to the atmosphere. However, we

Table 2.5 Global Emissions of Carbon, Sulfur, and Nitrogen Species due to Industry, Mining, and Natural Gas Transmission[a]

Species	Emission[b]	Sources
CO_2-C	150	Cement industry
CO-C	56	Iron and steel industry
SO_2-S	~10	Smelting, petroleum refining, H_2SO_4 production
NO_x-N	1.2	Petroleum refining, HNO_3 industry
HC	23	2/3 from the use of organic solvents
CH_4-C	26	Coal mining
CH_4-C	45	Gas drilling, transmission

[a] For references see the text.
[b] Values given in Tg per year.

discuss briefly in the following only the emission of substances relevant to the scope of this book (see Table 2.5).

As it was previously mentioned, during cement manufacturing a certain amount of carbon dioxide is liberated. In 1988 this amount was equal to 150 Tg C yr^{-1}, which can be neglected compared to total anthropogenic emission. In contrast, the industrial carbon monoxide emission is relatively important (Logan et al., 1981). It is around 56 Tg C yr^{-1}. This emission is mostly caused by pig iron and steel industry.

Among toxic metals emitted during fossil fuel combustion, vanadium, zinc,* nickel, and copper must be first of all mentioned. The combustion of fossil fuels causes 98% of global anthropogenic vanadium emission (86 Gg yr^{-1}) (see Nriagu and Pacyna, 1988). The corresponding figure for nickel is 73% of the total (56 Gg yr^{-1}). The emission of the other two metals mentioned is dominated by industrial processes. On the other hand, lead—this notorious air pollutant—is released mainly in automobile exhausts because of the lead content of gasoline. Global lead emission due to oil and gasoline combustion is estimated to be 259 Gg yr^{-1} (Nriagu and Pacyna, 1988), which is 78% of the total strength of anthropogenic lead sources. However, it should be taken into account that these data refer to 1983. Since that year lead emission has certainly decreased because of the decrease or stop in the use of lead in gasoline.

* In the case of zinc the emission of wood burning is very important.

During smelting, petroleum refining, and industrial sulfuric acid production, sulfur dioxide is emitted to the atmosphere. According to Cullis and Hirschler (1980), industrial processes produce about 15% of the total anthropogenic SO_2 emission. Taking into account the amount released during fossil fuel burning, this means that the total anthropogenic emission is around 70 Tg S yr^{-1}, a value used recently by Langner and Rodhe (1991) in their model calculations.

On the basis of the emission structure in the U.S. Logan (1983) concludes that the global industrial nitrogen oxide emission is controlled by petroleum refining and manufacture of nitric acid and cement. In this way she obtains for the global source strength a value of 1.2 Tg N yr^{-1}, which is not too large considering the emissions from the soil microbial activity and fossil fuel burning.

Among anthropogenic sources liberating hydrocarbons (NMHC) to the atmosphere the use of organic solvents is of particular significance; it amounts to 15 Tg yr^{-1} (Ehhalt et al., 1986). Further, a total amount of 8 Tg is emitted annually by the chemical industry, petroleum refineries, as well as during oil and gas production.

Coal and lignite mining provides a methane source, since coals contain a certain amount of this gas. This emission term is estimated to be 26 Tg C yr^{-1} globally with a range of 14 to 38 Tg C yr^{-1} (Cicerone and Oremland, 1988). According to the same authors an even larger quantity (45 Tg C yr^{-1}) is released during natural gas drilling, venting, and transmission.

Since the beginning of the thirties different *chlorofluorocarbons* (CFCs) have been used for several purposes. The chlorofluoromethanes like CCl_3F (CFC-11) and CCl_2F_2 (CFC-12) serve as refrigerants, propellants, and inflating agents in the manufacture of foam materials. On the other hand $C_2Cl_3F_2$ (CFC-113), carbon tetrachloride (CCl_4), and methylchloroform (CH_3CCl_3) are used widely as solvents. While methylchloroform reacts with OH radicals, the other species are chemically inert in the troposphere (their residence time is more than 50 years) and enter the stratosphere, where they are destroyed under the influence of ultraviolet solar radiation. It should be noted that halocarbons containing bromine like $CBrClF_2$ (halon 1211) and $CBrF_3$ (halon 1301) are also applied by man as fire extinguishers. However, their atmospheric importance is much less than that of chlorofluoromethanes.

Table 2.6 Annual Emission Rate of Some Chlorofluorocarbons in Tg per Year[a]

Species	Emission	Use
CFC-11	0.35	Refrigerant, propellant
CFC-12	0.45	Refrigerant, propellant
CFC-113	0.15	Solvent

[a] Data taken from IPCC, 1990.

Table 2.7 Global Industrial Emission of Some Trace Metals in Gg per Year[a]

Metal	Industry	Incineration
Cd	5.7	0.9
Cu	25	1.6
Ni	13	0.3
Pb	65	2.4
V	0.8	1.2
Zn	105	5.9

Note: For each metal a wide emission range is given by the authors referenced. The data in this table gives median values calculated from the range published

[a] The table also gives the emissions in the same unit due to refuse incineration (data taken from Nriagu and Pacyna, 1988).

The emission of halocarbons increased continuously until about 1975. It has decreased due to the efforts made to mitigate their effects on the stratospheric ozone layer. The present emission rates of some species is given in Table 2.6 according to the report of IPCC (1990).

Industry, mostly metallurgy, releases to the atmosphere a lot of metals. Table 2.7 summarizes the information available for some selected elements. For zinc, copper, and cadmium the ratio of industrial emission to the total strength of anthropogenic sources is high in particular. For Zn, Cu, and Cd this ratio is 80, 69, and 75%, respectively.

2.4.3 Incineration and Decay of Waste Materials

Human activities result in a large amount of solid and liquid waste. A part of this waste is burned while another part is disposed of in landfills.

The waste incineration produces a certain amount of carbon

dioxide, as well as sulfur and nitrogen oxides. However, the global emissions due to waste incineration are relatively insignificant compared to other anthropogenic source terms. Thus, for carbon dioxide the burning of different wastes gives only 3.5% of the total man-made emission (Hirschler, 1981). The corresponding figures for global SO_2 emission (Cullis and Hirschler, 1980) and U.S. NO_x release (Finlayson-Pitts and Pitts, 1986) are 0.3 and 0.4%, respectively.

In contrast to this, an important amount of methane is produced in landfills during the decay of organic matters. It is estimated (Bingemer and Crutzen, 1987; Cicerone and Oremland, 1988) that the global methane emission in carbon equivalent varies between 15 and 50 Tg yr^{-1} with a most probable value of 30 Tg yr^{-1}. One can assume (Bingemer and Crutzen, 1987) that the same amount of carbon dioxide is also liberated from landfills.

The release of carbon monoxide to the atmosphere due to waste disposal and treatment is relatively important. According to Logan et al. (1981) the global figure is equal to 8.5 Tg C yr^{-1}.

Waste incineration also contributes to anthropogenic metal emissions. As the last column of Table 2.7 shows, this source term for various metals has a magnitude of 0.1 to 1.0 Gg yr^{-1}, which is not too large compared to the intensity of other sources.

2.5 AEROSOL PARTICLE RELEASE BY OCEANIC AND CONTINENTAL SURFACES

2.5.1 Sea Salt Particles

The oceans provide an important atmospheric source of primary aerosol particles. While their concentration is low (~1 cm^{-3}) compared to the total particle number (see Section 3.1), due to their large size the sea salt particles control the aerosol mass burden in oceanic environment and play a role in precipitation formation and other atmospheric processes.

Sea salt particles can be formed by direct dispersal of ocean water from the foam of the waves. However, these particles are generally too large to remain airborne, even after water evaporation. A much greater number of particles is produced by the bursting of gas bubbles reaching the water surface.

According to the laboratory work of Moore and Mason (1954) this process takes place in two stages. In the first stage, when the bubble arrives to the surface, small particles are ejected from the bursting water film. In the second stage, a thin jet is formed by the water flowing into the cavity remaining in the surface after the rupture. The particles formed in the second stage are less numerous and their sizes are in the giant range (Woodcock, 1953).

The sea salt particles produced in this way are composed mostly of sodium chloride, which reflects the composition of sea water. Among other substances, marine particulate matter also contains a large amount of sulfates. Furthermore, during their rise through the water bubbles scavenge surface-active organic materials, which are partly injected into the air when the bubbles burst.

Woodcock (1953), as well as Moore and Mason (1954), originally demonstrated that the rate of bubble formation increases with increasing wind speed. In a more recent work A. Mészáros and Vissy (1974) reported that over the oceans the correlation between the number of sea salt particles and the wind speed becomes gradually weaker as the particle size decreases. Thus, the smallest sea salt particles ($r < 0.3$ μm) may originate from a type of bubble the formation of which is independent of the wind speed.

The relation between the bubble size and the number of airborne particles produced upon bursting was studied by Day (1963) during his laboratory investigation. He pointed out that the number of particles increases with increasing bubble size. A bubble with a size of several millimeters forms some hundreds of particles when it bursts. The results of simultaneous observations of the size spectra of bubbles in foam patches and giant sea salt particles in the air over a surf zone in Texas show that both spectra follow the gamma distribution function (Podzimek, 1984). On the basis of his atmospheric observations made in Hawaii, Blanchard (1969) assumed that the intensity of sea salt particle formation is between 25 and 100 $cm^{-2}\ s^{-1}$ at the surface of the ocean. This range agrees with the laboratory results of Moore and Mason (1954). On a mass basis 1000 to 2000 Tg of sea salt is released annually to the air. If the mass of very large particles returning very quickly from the air to the ocean is omitted from the calculation a value of 300 Tg yr^{-1} is obtained (SMIC, 1971).

Considering the average composition of sea water 7.5% of this amount is composed of sulfates. On the other hand, atmospheric observations carried out by Cachier et al. (1986) show that the organic carbon emission is probably <1% of the total sea salt mass emitted. Sea salt particles also contain different microelements. It can be estimated (Nriagu, 1989) that the median value of the emission of different metals has a magnitude of 1 Gg yr^{-1} or less.

Airborne sea salt particles are transported to higher levels and over the continents by atmospheric motions. Thus, such particles were observed over Australia (Twomey, 1955), North America (Byers et al., 1957), and Europe (Mészáros, 1964). Due to removal processes near the surface level, as well as to convective motions, the concentration of sea salt particles increases with increasing height in the lower troposphere over the continents, in contrast to the situation over the ocean where their number decreases with increasing altitude (Woodcock, 1953; Lodge, 1955; Blanchard et al., 1984).

Except for cases when the sea surface is covered by petroleum products, human activities do not influence the emission of sea salt particles. However, their composition can be modified since gaseous sulfure and nitrogen oxides, partly of anthropogenic origin, react with sea salt to form sodium sulfate and nitrate. This reaction liberates a relatively large amount of hydrochloric acid vapor to the atmosphere. The anthropogenic portion of this source is an open question.

2.5.2 Particles of Soil and Mineral Origin

An important amount of coarse (particle radius larger than ~1 μm) dust particles originates from the solid surface of the Earth due to mechanical disintegration. This dispersal is obviously due to the effect of wind on rocks and soils. A well-known and highly visible example of this process is the formation of dust clouds and storms. However, the quantitative explanation of this particle production mechanism is not easy, except when some external mechanical force agitates the surface (vehicles, animals, people, etc.). The main reason for the difficulties in the explanation is the decrease of the wind speed with decreasing height above surface, usually extrapolating to zero wind speed at the surface. It is believed that turbulent flow is necessary (see Twomey, 1977) for the detachment of grains. According to the most acceptable estimates, the global strength of this source is 100 to 500 Tg yr^{-1} (SMIC, 1971; Prospero, 1984). Obviously a fraction of this emission is of anthropogenic origin due to land use practice and vegetation cutting. However, the magnitude of man's contributions is not well determined at present.

The composition of dust emission can be characterized by determining the so-called aerosol-crust enrichment factor, which is based on the analyses of atmospheric aerosol samples and rocks or soils. The enrichment factor (EF) defined by Rahn (1976) is given by the following expression

$$EF = \frac{(X/Ref)_{aerosol}}{(X/Ref)_{crust}}$$

where X is the concentration of the element considered in the aerosol or the crust and "Ref" is the concentration of a crustal reference element (generally Al, Ti, or Fe). It follows from this definition that a certain element is considered to be of mineral origin if its enrichment factor is near unity. By using this approach and Ti as a reference element A. Mészáros et al. (1984) demonstrated that, in agreement with the results of many other workers, under continental conditions Al and Si are the most important elements of soil origin; these elements almost certainly occur in silicate compounds. A relatively large amount of iron, sodium, calcium, and magnesium can also be found in aerosol samples under continental conditions. The calculations of Nriagu (1989) show that in a much smaller amount different metals are also released to the atmosphere during soil and rock disintegration. The emission rate of elements like copper, nickel, lead, vanadium, and zinc is between 4 and 20 Gg yr^{-1}. Lead analyses of samples of Antarctic ice taken from various depths (time in the past) indicate that between 155,000 to 26,000 years ago Pb in the air was of soil dust origin with some volcanic contribution (Boutron et al., 1987).

Some particles of crustal origin are removed from the air in the vicinity of sources, while another fraction is transported to great distances. Thus, particulate matter collected over the Atlantic Ocean contains a significant quantity of Saharan dust under some conditions (Junge and Jaenicke, 1971). In fact, such

Table 2.8 Comparison of Natural and Anthropogenic Surface Source Strength of Different Trace Gases[a]

Gas	Natural Biogenic	Other Natural	Anthropogenic	Anthropogenic per Total (%)
CO_2	100	—	7	6.5
CO	75	—	505	87
CH_4	121	—	276	69
N_2O	6	—	0.4	6.3
DMS/SO_2	25	8	76	70
NO_x (x=1,2)	10	—	28	74
NH_3	22	—	26	54
NMHC	765	—	90	11

Note: For emissions given by a range of values the averages are tabulated. For references see the appropriate parts of the text.

[a] Terrestrial carbon dioxide emissions given in Pg of carbon per year, while other values expressed in Tg of C, N, S, and NMHC per year.

dust particles were collected and identified even over the West Indies (Prospero, 1968; Blifford, 1970). Furthermore, dust particles from Asian desert were identified by several workers (see Braaten and Cahill, 1986) at the Mauna Loa Observatory in Hawaii. Elements of soil origin like Al and Fe were even found in relatively high concentrations at the South Pole, Antarctica (Zoller et al., 1979). It was also shown (Andreae et al., 1986) that, during their transport over the oceans, silicate species can be mixed with sea salt components.

2.6 STRENGTH COMPARISON OF DIFFERENT SOURCES AT THE EARTH'S SURFACE

2.6.1 Gaseous Compounds

The source strength of the emission of different gaseous compounds discussed in this chapter is summarized in Table 2.8. In the table anthropogenic terms contain biogenic emissions owing to human activities (e.g., biomass burning, cattle husbandry, agriculture). The column "other natural" gives the emission due to volcanic activity (sulfur). In Table 2.8 the release of chlorofluorocarbons is not given, since the emission of the species in Table 2.6 is entirely man-made. It also should be noted that for the natural biospheric source of CO_2 the value published in IPCC (1990) is used, it is lower by 12% than the sum of natural biogenic emissions in Figure 2.2.

Data tabulated show that in absolute terms man produces the highest anthropogenic emission in the case of carbon dioxide. Although in relative units human contribution is low compared to the strength of natural biogenic CO_2 sources, the carbon cycle in nature is so delicate that natural sinks are not able to remove man-made excess. For this reason CO_2 concentration in the atmosphere increases continuously (see Section 2.1).

Both in absolute and relative terms human activities contribute significantly to carbon monoxide and methane emissions. For CO this is caused by the high amount released during biomass burning (see Table 2.3), while for methane more than the half of anthropogenic emissions is due to enteric fermentation of domestic animals and rice production (see Table 2.2). In the case of CO it also should be noted that an important part of this gas forms in the air by chemical reactions (see Section 3.3), which is not included in the table. In the case of nitrous oxide the man-made contribution is less than 10%. The present rise of its concentration is also moderate relative to the increase in carbon dioxide and methane levels (see Section 2.1).

The anthropogenic contribution of sulfur and nitrogen compounds to the global emission is also very high. In addition to data given in Table 2.8 the man-made fraction of the emissions of S and N species in the air over industrialized regions is significant in particular. Thus, over Europe (Mészáros and Várhelyi, 1982) and North America (Galloway and Whelpdale, 1980) only 4% of the sulfur emission is of natural origin. For nitrogen oxides and ammonia this fraction over Europe lies between 5 and 20% as discussed by Bónis et al. (1980).

In contrast to sulfur and nitrogen species, only a small fraction of the global hydrocarbons emission is due to human activities. Even over populated areas the relative importance of man-made sources is not as high as for S and N containing gases. Over the U.S. the anthropogenic hydrocarbons emission during summertime is only the third of the strength of natural sources (Fehsenfeld, 1991). In central Eastern Europe (Molnár,

Table 2.9 Natural vs Anthropogenic Emissions of Trace Metals to the Atmosphere in 1983[a]

Metal	Anthropogenic[b]	Natural[b]	Anthropogenic per Total (%)
Cd	7.6	1.3	85%
Cu	35	28	64%
Ni	56	30	65%
Pb	332	12	96%
V	86	28	75%
Zn	132	45	66%

[a] Nriagu, 1989.
[b] Values expressed in Gg per year.

1990) and Western Europe (CONCAWE, 1986) annual emissions from natural and anthropogenic sources are comparable due to the hydrocarbon release from forests.

2.6.2 Aerosol Particles

The emission of primary aerosol particles from natural and anthropogenic sources is not well understood. For this reason the comparison of the strength of natural and man-made sources is not an easy task. In spite of the difficulties involved such comparison was made for different metals. In Table 2.9 the data given by Nriagu (1989) are summarized for selected elements. One can see that mankind has become the dominant agent in the control of the atmospheric cycle of toxic metals. Human contribution is high in particular for lead and cadmium due to gasoline combustion and metallurgical activities, respectively.

Finally we note that an important part of atmospheric aerosol particles is not emitted directly to the atmosphere. Smaller particles form generally by gas-to-particle conversion in the air as presented in Section 3.4 dealing with aerosol formation by chemical reactions and condensation.

3

Formation and Destruction of Trace Substances within the Atmosphere

3.1 INTRODUCTORY REMARKS: ELEMENTS OF PHOTOCHEMISTRY AND REACTION KINETICS

3.1.1 General Considerations

In the atmosphere many chemical reactions occur. Chemical reactions remove certain species from the air by producing new components. In this way they provide either a source or a sink of the compounds involved. Homogeneous gaseous reactions take part in gas phase, while in heterogeneous processes other phases (e.g., aerosol particles, cloud elements) also play a role. Since air is an oxidative medium, chemical reactions produce species in more oxidative states, e.g., ozone from molecular oxygen. In certain cases gaseous compounds with low saturation vapor pressure are formed (e.g., sulfuric acid, nitric acid, organic vapors). Consequently they condense in the air together with water vapor to create minute liquid aerosol particles.

Chemical reactions in the atmosphere can be divided into two categories. The energy of *photochemical* reactions is provided by solar radiation, while the rate of *thermal* reactions is determined by the thermal energy (temperature).

The principle of photochemical processes can be summarized as follows. A certain atmospheric gaseous component, A, absorbs a given band in the UV or visible solar spectrum. Due to the energy of the absorbed photon, hν (where h is Planck's constant and ν is the frequency of the radiation), A is changed to an excited state. Because of this energy, the molecule decomposes or reacts with another compound, B, (disregarding quenching and fluorescence phenomena). The process may be symbolized in the following way (Leighton, 1961):

$$A + h\nu \rightarrow A^* \qquad \text{absorption*}$$
$$A^* \rightarrow D_1 + D_2 + \ldots \qquad \text{dissociation}$$
$$A^* + B \rightarrow D_1 + \ldots \qquad \text{direct reaction}$$

The last two steps in this sequence are termed primary photochemical processes. If the atmospheric concentration of the species is denoted by square brackets then

$$\frac{d[A^*]}{dt} = k_a[A]$$

where k_a is the absorption rate of photons. The rate of the primary photochemical process will be:

$$\frac{d[D_1]}{dt} = k_a\Phi[A]$$

where Φ is the so-called quantum yield. Its value is equal to the ratio of the number of A reacted to that of excited molecules.

The primary photochemical processes are generally followed by secondary thermal reactions the energy of which is provided by the thermal agitation of molecules. In the case of a bimolecular reaction**

$$D_1 + E \rightarrow X + Y$$

$$-\frac{d[D_1]}{dt} = \frac{d[X]}{dt} = k[D_1][E]$$

* Asterisk denotes the excited state.

** In these equations D_1, E, X, and Y each denote an appropriate molecule.

where k is the rate constant, which is equal to the quantity of D_1 reacted per unit time if the $[D_1]$ and [E] concentrations are also unity. It can be seen from the last equation that the loss rate of D_1 is equal to the formation rate of X in molar units.

The above bimolecular reaction is said to be second order, since its rate depends on the product of two concentrations. Generally, the order of a reaction is the sum of the exponents of concentrations on the right-hand side of kinetic equations. Thus, the primary photochemical reaction discussed above is a first order process since its rate depends only on the concentration of A. In the case of photochemical reactions the rate constant is given by the product of the absorption rate and quantum yield.

3.1.2 Examples of Reaction Kinetics

Both in tropospheric and stratospheric chemistry nitrogen dioxide plays an important role. NO_2 can be destroyed by visible solar radiation to form nitric oxide and atomic oxygen. This is a very important reaction since it produces atomic oxygen essential for ozone formation. On the other hand, nitric oxide formed by the photolysis of NO_2 reacts with ozone to recreate nitrogen dioxide:

$$NO_2 + h\nu \rightarrow NO + O \tag{3.1}$$

$$NO + O_3 \rightarrow NO_2 + O_2 \tag{3.2}$$

The kinetic rates of the above reactions are

$$-\frac{d[NO_2]}{dt} = k_a \Phi [NO_2] \tag{3.1a}$$

$$\frac{d[NO_2]}{dt} = k_2 [NO][O_3] \tag{3.2a}$$

where $k_1 = k_a\Phi = 8 \times 10^{-3}\ s^{-1}$ if $\lambda > 310$ nm, while at 298 K $k_2 = 1.8 \times 10^{-14}\ cm^3\ molecule^{-1}$ (Warneck, 1988). Using these figures the equilibrium concentration of NO_2, NO, or O_3 can be calculated, since in equilibrium the left-hand side of both (3.1a) and (3.2a) is equal to zero. Thus, the equilibrium ozone concentration is

$$[O_3] = \frac{k_a \Phi [NO_2]}{k_2 [NO]} \tag{3.3}$$

It follows from the last equation that the ozone concentration is high when the $[NO_2]/[NO]$ ratio is high. In contrast, when NO is significant compared to NO_2, nitric oxide consumes ozone molecules.

Hydroxyl free radicals (OH) are key species in atmospheric chemistry (Crutzen, 1974; Logan et al., 1981). They are created by the interaction of excited oxygen atoms that are formed by ozone photodissociation and water vapor with a rate constant of 2.2×10^{-10} cm^3 $molecule^{-1}$ s^{-1} (at a temperature of 298 K) (Warneck, 1988):

$$O^* + H_2O \rightarrow 2OH \tag{3.4}$$

On the other hand, they are removed mainly by the reaction with carbon monoxide and methane:

$$OH + CO \rightarrow CO_2 + H \tag{3.5}$$

$$OH + CH_4 \rightarrow CH_3 + H_2O \tag{3.6}$$

At a temperature of 298 K the rate constants of Reactions 3.5 and 3.6 are $k_5 = 2.8 \times 10^{-13}$ cm^3 $molecule^{-1}$ s^{-1} and $k_6 = 7.7 \times 10^{-15}$ cm^3 $molecule^{-1}$ s^{-1}, respectively, (Warneck, 1988). The equilibrium OH concentration is

$$[OH] = \frac{2k_4 [O^*] [H_2O]}{k_5 [CO] + k_6 [CH_4]} \tag{3.7}$$

By applying the above rate constants one calculates equilibrium OH concentrations around 10^6 cm^{-3} as a function of the concentration values used. It follows from Equation 3.7 and the appropriate rate constants that, if the concentration of CO increases in the atmosphere, the methane level also increases, OH is consumed by the faster reaction with CO.

Similar kinetic calculations can be carried out for any case, even if the number of the reactions is high. Thus, kinetic models make it possible to calculate the variations of the level of any constituent as well as its equilibrium concentration. Such models are widely used in studies of atmospheric chemistry, including concentration changes due to anthropogenic emissions.

3.2 FORMATION AND DESTRUCTION OF STRATOSPHERIC OZONE

3.2.1 Natural Homogeneous Chemistry

The first kinetic model for ozone formation and destruction in the stratosphere was proposed by Chapman (1930). According to his theory only oxygen takes part in the O_3 cycle. Thus, molecular oxygen absorbs solar radiation with wavelengths between 0.18 and 0.21 μm. Excited molecules dissociate into two atomic oxygens which combine with O_2 to form ozone by a secondary thermal reaction:

$$O_2 + h\nu \rightarrow O + O \tag{3.8}$$

$$O_2 + O + M \rightarrow O_3 + M \tag{3.9}$$

where M is a third body, generally nitrogen. Ozone strongly absorbs in the 0.20 to 0.32 μm band of the UV part of the solar spectrum and in the visible range between 0.45 and 0.70 μm wavelengths. Absorption in the UV band screens out lethal UV radiation from the solar spectrum and results in the heating of the air layer between 15 to 50 km.

O_3 molecules in the stratosphere are decayed by photodissociation owing to radiation absorption:

$$O_3 + h\nu \rightarrow O_2 + O \tag{3.10}$$

or by the thermal reaction between ozone and atomic oxygen:

$$O_3 + O \rightarrow 2O_2 \tag{3.11}$$

If the rate constants of the photochemical reactions (Reactions 3.8 and 3.10) are denoted by j_8 and j_{10} and those of the thermal reactions (Reactions 3.9 and 3.11) by k_9 and k_{11}, then in equilibrium

$$\frac{d[O_3]}{dt} = 0 = k_9[O_2][O][M] - k_{11}[O_3][O] - j_{10}[O_3] \quad (3.12)$$

$$\frac{d[O]}{dt} = 0 = 2j_8[O_2] - k_{11}[O_3][O] - k_9[O_2][O][M] + j_{10}[O_3] \quad (3.13)$$

The combination of Reactions 3.12 and 3.13 yields an equilibrium equation of second order which can be solved for O_3 by using information on the distribution of solar radiation and temperature. Applying rate constants available before about 1960, fairly good agreement between theory and observation was obtained if atmospheric transport in the lower stratosphere was taken into account (see Junge, 1963). Moreover, the height of the maximum O_3 concentration was found to be nearly independent of the assumptions used (~23 km) and in agreement with O_3 vertical profile measurements.

More recent studies, however, indicated that older rate constants used in ozone calculations were incorrect. Thus, newer theoretical results showed that ozone concentrations are grossly overestimated by the Chapman's model compared to satellite and rocket observations (Hunt, 1966). This raised the possibility that trace gases coming from the troposphere play an important role in stratospheric chemistry. First, it was proposed that, aside from the ozone removal summarized above, hydrogen containing free radicals like OH formed from water vapor of tropospheric origin also participate in ozone destruction (Hunt, 1966):

$$OH + O_3 \rightarrow HO_2 + O_2 \quad (3.14)$$

$$HO_2 + O_3 \rightarrow OH + 2O_2 \quad (3.15)$$

$$\text{net } O_3 + O_3 \rightarrow 3O_2 \quad (3.15a)$$

It was later demonstrated that in the lower stratosphere Reaction 3.15 is a relatively slow process due to the competitive reaction of NO with HO_2:

$$HO_2 + NO \rightarrow NO_2 + OH \quad (3.16)$$

which is followed by NO_2 photolysis and ozone formation. This latter process regenerates the ozone (see Reactions 3.1 and 3.9) quantity consumed by Reaction 3.14. Model calculations made by different research workers (Johnston and Podolske, 1978; Wofsy and Logan, 1982) show that these processes account for only 5 to 10% of O_3 destruction, while about 10 to 20% of ozone molecules are removed by Chapman's reactions (Reactions 3.10 and 3.11).

However, in the upper stratosphere OH radicals also react with atomic oxygen, which leads to the following chain reactions:

$$O + OH \rightarrow O_2 + H \quad (3.17)$$

$$H + O_3 \rightarrow O_2 + OH \quad (3.18)$$

$$H + O_2 + M \rightarrow HO_2 + M \quad (3.19)$$

$$O + HO_2 \rightarrow OH + O_2 \quad (3.20)$$

which is terminated by the combination of OH and HO_2 (hydroperoxyl radical) to form water and molecular oxygen. Wofsy and Logan (1982) calculate that the above chain reactions are responsible for ~15% of the ozone loss.

At the beginning of the seventies it was speculated (Crutzen, 1971) that nitric oxide molecules play a key role in stratospheric ozone destruction:

$$NO + O_3 \rightarrow NO_2 + O_2 \quad (3.2)$$

$$NO_2 + O \rightarrow NO + O_2 \quad (3.21)$$

Nitric oxide formation in the lower stratosphere is initiated by the photodissociation of nitrous oxide of surface origin. This produces N_2 and excited oxygen atoms. Oxygen atoms then react with N_2O to form NO. On the basis of kinetic model calculations it is estimated (see the above references) that Reactions 3.2 and 3.21 are very significant in ozone removal;

they remove at least one third of ozone molecules from the stratosphere. It should be noted that, owing to their energy, solar protons can also create nitric oxide in the upper stratosphere at high latitudes. Under special conditions the ozone depletion by NO formed in this way can be rather significant (Stephenson and Scourfield, 1991).

Finally, it was proposed that *chlorine* is also an important element in ozone chemistry, as discussed by Stolarski and Cicerone (1974) and Molina and Rowland (1974). Under natural conditions chlorine atoms form in the stratosphere by the photolysis of methyl chloride emitted mainly from biological processes occurring in the oceans (see Section 2.2.2). CH_3Cl molecules reach the stratosphere where they are dissociated to form CH_3 and Cl, leading to the following reaction sequence:

$$Cl + O_3 \rightarrow ClO + O_2 \quad (3.22)$$

$$ClO + O \rightarrow Cl + O_2 \quad (3.23)$$

$$\text{net} \quad O_3 + O \rightarrow 2O_3 \quad (3.23a)$$

It should be noted that the destruction of ozone by nitrogen oxides and chlorine/chlorine oxide is not an infinitive process, in spite of the fact that initial compounds are reformed during ozone removal. This is caused by the fact that these species also react with other gaseous components to create so-called "reservoir" molecules, ineffective in ozone chemistry. Thus, chlorine atoms react with methane, while NO, NO_2, and ClO are removed by their reactions with OH radicals. These reactions result in the formation of hydrochloric acid, nitrous acid, and nitric acid, which are chemically rather stable and slowly transported into the troposphere to leave the atmosphere by deposition. Even two reactive species, ClO and NO_2, can be combined to form $ClNO_3$ (chlorine nitrate) also inert in the ozone cycle.

3.2.2 Anthropogenic Effects: Heterogeneous Chemistry

It follows from the theory of ozone destruction that human activities may constitute a threat to stratospheric O_3. This possibility was first stressed by Crutzen (1971) and Johnston (1971) by stating that supersonic aircraft flying in the stratosphere could emit nitrogen oxides in a quantity sufficient to alter chemical processes. However, it was found that supersonic transport for the future is not predicted to have a significant effect on stratospheric chemistry (see, e.g., WMO, 1976). Further, reevaluation of the original concept (Johnston and Podolske, 1978) makes it clear that nitrogen oxides destroy ozone at high altitudes but form it at low altitudes through NO_2 dissociation (see Reactions 3.1 and 3.9).

In spite of the fact that it was also proposed that the use of fertilizers producing N_2O can affect the ozone layer, at the end of the 1970s it was concluded (NAS, 1979) that the release of CFCs appears to be the greatest and most immediate threat to the ozone shield.* This is caused partly by the common use of these materials and partly by the relatively low natural halocarbon concentrations in the air. Thus, in the troposphere the concentration of methyl chloride, the most important chlorine containing natural compound, is only around 0.5 ppb (Penkett et al., 1980), which is exceeded by the level of CFCs of anthropogenic origin (see Table 2.1). In contrast, the concentration of N_2O of biological origin is high (~0.4 ppm)—actually much higher than man-made nitrogen compounds. Chlorofluorocarbons, chemically inert in the troposphere, are photochemically destroyed in the stratosphere by solar radiation with wavelengths shorter than 0.23 μm to form chlorine atoms, e.g.:

$$CCl_2F_2 + h\nu \rightarrow CClF_2 + Cl \qquad (3.24)$$

which is followed by the catalytic reaction chain discussed above (Reactions 3.22 and 3.23). This possibility was first raised by Molina and Rowland (1974) to explain the atmospheric fate of CFC molecules. After this proposition many model calculations were carried out to estimate human effects on the stratospheric ozone layer. The results obtained varied as a function of the assumptions used. Finally, the discovery of the

* Note that in a more recent paper Johnston et al. (1989) proposed a relatively large ozone reduction due to NO injection by high speed aircraft in the stratosphere. However, Bekki et al. (1991) argue that the reduction becomes much less if heterogeneous chemistry is also considered in model calculations.

spring ozone hole over Antarctica (Farman et al., 1985) changed considerably our concept concerning anthropogenic destruction of ozone in the stratosphere. For the explanation of the phenomenon it was necessary to assume that heterogeneous chemical processes are able to reform active species from reservoir species containing man-made chlorine. The explanation was based on laboratory experiments showing that ice crystals formed at very low temperatures (e.g., in the polar stratosphere during wintertime) absorb both hydrochloric acid (HCl) and chlorine nitrate ($ClNO_3$). In the solid phase these two reservoir species combine:

$$ClNO_3 + HCl \rightarrow Cl_2 + HNO_3(H_2O) \tag{3.25}$$

Their combining liberates chlorine molecules (Cl_2) in gaseous form and creates nitric acid trihydrate, which remains in the solid phase. The efficiency of the process is a function of the temperature and chlorine concentration. If Cl_2 molecules are photochemically destroyed in the spring under the effect of solar radiation the Cl atoms react immediately with ozone as discussed above. Ozone removal is further promoted by the formation of $(ClO)_2$ dimer from two ClO molecules. This formation process is possible if the ClO concentration is high (Barett et al., 1988). $(ClO)_2$ is photodissociated to also create atomic chlorine:

$$(ClO)_2 + h\nu \rightarrow Cl + ClO_2 \tag{3.26}$$

In spite of the fact that bromine also can participate in the process, Reactions 3.25 and 3.26 can explain *in principle* the formation of the ozone hole as it will be discussed further in Section 6.3. We only note here that it was proposed recently that sulfate particles, formed in the stratosphere after volcanic eruptions, can also influence sporadically the stratospheric ozone layer similar to ice crystals in polar stratospheric clouds (Hofmann and Solomon, 1989). Another possibility is that N_2O_5 formed by the oxidation of nitrogen oxides interacts with sulfuric acid solution droplets to form nitric acid vapor (Rodriguez et al., 1991). This reduces the concentration of nitrogen oxides, which slows down the removal of ClO molecules.

3.2.3 Atmospheric Observations

The atmospheric ozone layer is monitored classically with ground-based optical devices called spectrophotometers. These instruments detect the intensity of solar radiation in two narrow wavelength bands. In one of these bands ozone strongly absorbs the radiation, while in the other O_3 has little absorption. This procedure makes it possible to determine a parameter termed the "total ozone". The unit of total ozone is cm, which represents the thickness of the layer that the same amount of ozone would form if it were separated from the air and held at normal temperature and pressure. Although such an observation gives the total ozone content in an air column, it represents practically the stratospheric ozone layer, since ~90% of O_3 molecules are observed in the stratosphere. Thus, Junge (1962) calculated the total ozone quantity in the atmosphere over the Northern Hemisphere and obtained a value of 1750 Tg. The corresponding figure for the tropospheric reservoir was found to be 130 Tg.

In the 1920s and 30s the total ozone was already measured at several stations using the spectrophotometers developed by Dobson.* Since the vertical profile of ozone can also be determined by spectrophotometric observations, these early measurements showed that the maximum concentration is detected at a level of about 22 km. Another interesting finding was the latitude distribution of total ozone, which has been proved by many further measurements including satellite observations. According to this finding the total ozone increased from the equator toward higher latitudes in contrast to the photochemical theory. Moreover, measurements indicated that the maximum of total ozone occurred during wintertime and not during the summer when the intensity of solar radiation is more significant. Thus, it was recognized that ozone distribution in the lower stratosphere is determined more by transport than by photochemistry owing to the poleward ozone flux in the stratosphere. This circulation pattern is particularly strong in winter and spring months when stratospheric air moves downward over polar regions and upward over the tropics. Thus, under natural conditions undisturbed by human activities

* For the historical survey of ozone observations see Warneck (1988).

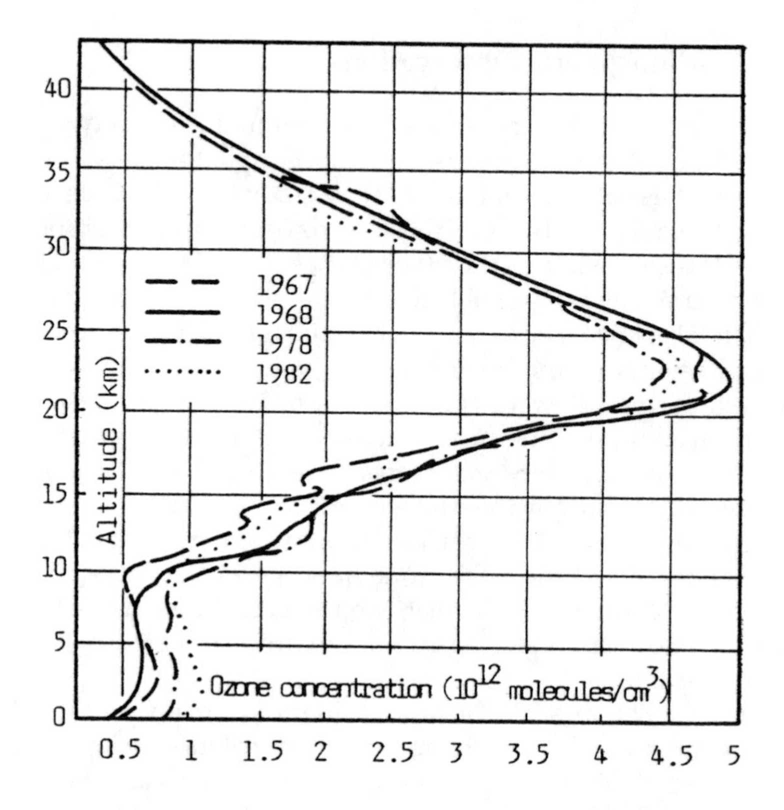

Figure 3.1 Vertical profile of ozone concentration in summer over Hohenpeissenberg, Germany (Brühl and Crutzen, 1989).

stratospheric dynamics lead to the accumulation of ozone rich air in the lower polar stratosphere.

The vertical profile of the ozone concentration can be measured more precisely by balloon-borne instruments. Figure 3.1 represents the results of such observations carried out at Hohenpeissenberg, Germany (Brühl and Crutzen, 1989). The curves indicate clearly that the majority of ozone is found in the stratosphere. Also, between 20 and 25 km an ozone concentration decrease with time can be observed. It is very probable that it is due to human activities—more exactly to the release of different chlorofluorocarbons to the atmosphere. On the other hand, observations reveal an ozone level rise in the troposphere. It is speculated that this is caused by emissions of different air pollutants as discussed in the next section.

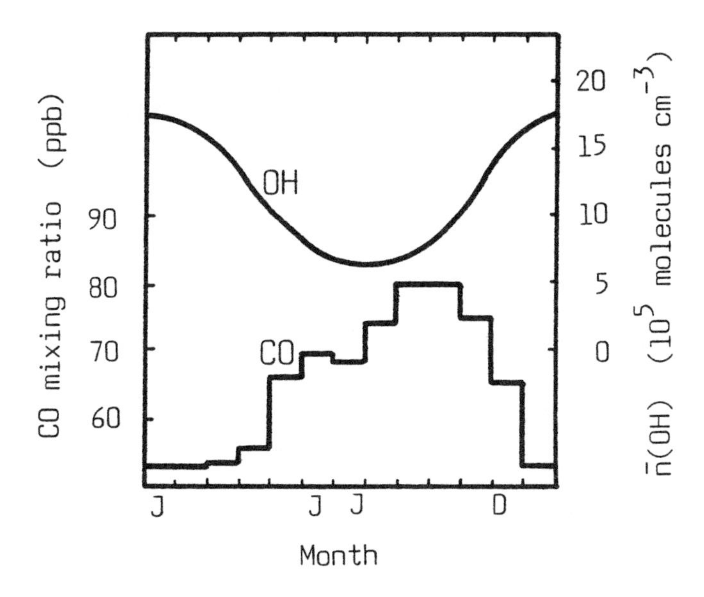

Figure 3.2 Total ozone trends as measured by the spectrometer on the Nimbus 7 satellite. The values are expressed in percent per year. The shaded area indicates when the trends are not statistically different from zero. "No date" areas refer to polar nights, where measurements are not possible (Stolarski et al., 1991).

In recent time the total ozone data have been carefully analyzed to look for possible trends that can be related to anthropogenic modifications. The results obtained by means of Dobson spectrophotometers at the surface have been critically evaluated by Bojkov et al. (1990), while total ozone trends deduced on the basis of satellite (Nimbus 7) observations have been studied by Stolarski et al. (1991). According to Bojkov and his associates data gained at the northern hemispheric stations located between 19° N and 64° N latitudes show a trend of –0.84% per decade between 1970 and 1986. The trend is significant especially in the winter at higher latitudes. Thus, it is –3% per decade during winter months at 55° N. An even higher decrease is indicated by satellite observation on the basis of measurements carried out between 1979 and 1990. Figure 3.2 represents the ozone decrease per year as published by Stolarski et al. (1991). One can see that around the Equator no changes are observed. Changes are important in both hemispheres in the winter near the pole. The decrease is significant in particular during spring months (September and October) at higher latitudes in the Southern Hemisphere in relation to the ozone hole detected over Antarctica during this part of the year (see Section 6.3).

3.3 HOMOGENEOUS TROPOSPHERIC CHEMISTRY

3.3.1 Basic Chemical Reactions: Tropospheric Ozone

As it was discussed in Chapter 2 natural and anthropogenic sources emit to the troposphere different gaseous species like methane, carbon dioxide, nitrogen oxides, and different hydrocarbons. In addition to these compounds ozone is transported to the troposphere from the stratosphere by slow diffusion and the tropopause* discontinuities. These gases, diluted in the air, initiate several gaseous reaction sequences (Logan et al., 1981) in which OH radicals (see Reaction 3.4) play an important part.

Among other things, tropospheric reactions result in the chemical removal of carbon dioxide and methane and the rapid oxidation of hydrocarbons and nitric oxide. The oxidation of nitric oxide produces nitrogen dioxide, which is photodissociated to form atomic oxygen necessary for ozone formation. Thus, it is speculated (Fishman and Crutzen, 1978) that an important part of O_3 in the troposphere is an *in situ* product. On the other hand, the oxidation of methane and other hydrocarbons leads to the formation of formaldehyde (CH_2O), molecular hydrogen,** and carbon monoxide.

The removal of carbon dioxide begins with Reaction 3.5:

$$CO + HO \rightarrow CO_2 + H \tag{3.5}$$

followed by the sequence

* The surface separating the troposphere from the stratosphere.

** H_2 can also be emitted by natural and anthropogenic sources at the surface. Recent studies show that its atmospheric level increased between 1985 and 1989 due to man-made emissions (Khalil and Rasmussen, 1990).

$$H + O_2 + M \rightarrow HO_2 + M \qquad (3.27)$$

$$HO_2 + NO \rightarrow NO_2 + OH \qquad (3.28)$$

The resulting species are carbon dioxide (this CO_2 source is unimportant relative to emissions at the surface), nitrogen dioxide, and OH radicals. However, if the NO concentration is low Reaction 3.27 is followed by Reaction 3.15. In the first case NO_2 and, consequently, O_3 are formed, while in the second O_3 is removed from the air. This means that NO_x concentrations are determinant concerning tropospheric ozone chemistry.

In the case of methane the reaction chain is as follows:

$$CH_4 + OH \rightarrow CH_3 + H_2O \qquad (3.6)$$

$$CH_3 + O_2 + M \rightarrow CH_3O_2 + M \qquad (3.29)$$

$$CH_3O_2 + NO \rightarrow CH_3O + NO_2 \qquad (3.30)$$

$$CH_3O + O_2 \rightarrow CH_2O + HO_2 \qquad (3.31)$$

The formaldehyde formed in this way dissociates under the influence of solar radiation with wavelengths 0.30 to 0.36 μm:

$$CH_2O + h\nu \rightarrow H_2 + CO \qquad (3.32)$$

It is to be noted that CH_4 can be substituted formally in these reactions by any other hydrocarbon. Since the process is more rapid for nonmethane hydrocarbons (NMHC) than for CH_4, the oxidation rate of NO molecules in the air is controlled by the concentration of NMHC molecules. Further, since the concentration ratio of NO_2 to NO determines whether ozone formation or removal occurs (see Section 3.1.2), the emissions of carbon monoxide and hydrocarbons control in an important way the tropospheric ozone chemistry. Ozone is a good indicator of photochemical smog episodes on local and regional scales (see Section 6.1). Thus one can conclude that man-made NO, NMHC, and CO molecules play a decisive role in the formation of secondary pollutants in the air.

According to Munn and Rodhe (1985) the annual production of ozone in the troposphere is 2100 Tg yr^{-1}, while the input from the stratosphere is 1000 Tg annually. The major part of O_3 formation occurs over the Northern Hemisphere due to the distribution of continents and anthropogenic sources. Two thirds of this ozone amount is removed from the air by chemical reactions,* while the balance is destroyed at the Earth's surface.

The photochemical activity of the air can also be characterized by another compound: the peroxyacetyl nitrate (PAN). Without chemical details we note that this species is formed by the oxidation of hydrocarbons in the presence of nitrogen oxides. PAN was first identified in photochemical smogs under polluted conditions. However, further studies revealed that it is an ubiquitous compound in the troposphere (Singh and Salas, 1983). In the free troposphere over the oceans its level can reach 0.1 ppb.

3.3.2 Removal Rate of Methane

The major part of methane molecules is removed from the air in the troposphere by Reaction 3.6, while a smaller part reaches the stratosphere. The rate constant of the reaction is 7.7×10^{-15} cm^3 $molecule^{-1}$ s^{-1} at a temperature of 298 K. One can speculate (see Warneck, 1988) that the average rate for the lower troposphere is 5.5×10^{-15} cm^3 $molecule^{-1}$ s^{-1} and the mean concentration of OH radicals is equal to 5×10^5 molecules cm^{-3}. It can be further calculated that the total mass (M) of methane molecules in the troposphere is 4177 Tg if the mean concentration (C: mixing ratio) is 1.72 ppm (see Table 2.1), since

$$M = C \frac{m_{methane}}{m_{air}} \times M_T$$

where $m_{methane}$ and m_{air} are the molecular weight of methane and air, respectively, and M_T is the total mass of tropospheric air (4.25×10^{21} g). Using the kinetic theory the methane loss is equal to

* Results of recent research show that a part of ozone removal by chemical reactions takes place in the clouds (Lelieveld and Crutzen, 1990).

$$\frac{dM}{dt} = k_6[OH]\,[M] \tag{3.33}$$

which must be multiplied by 3.15×10^7 to convert seconds to years. In this way Reaction 3.33 yields a methane removal rate of 329 Tg yr^{-1} (~250 Tg C yr^{-1}). This is comparable to the total emission of 397 Tg C yr^{-1} given in Table 2.8, if we take into account that at least 60 Tg CH_4 is transported annually to the stratosphere (Warneck, 1988).

3.3.3 Internal Sources and Sinks of Carbon Monoxide in the Atmosphere

Within the atmosphere two internal (chemical) carbon monoxide sources can be distinguished. The first source is provided by the formation of CO from methane molecules, while the second is the oxidation of other hydrocarbons (NMHC). The rate of these formation processes can be estimated by applying kinetic arguments and considering concentrations measured in the atmosphere. The results obtained vary as a function of the input data used. Thus, Logan et al. (1981) propose that the oxidation of methane produces globally 350 Tg CO–C yr^{-1}, while the oxidation of NMHC of natural origin results in a production rate of 240 Tg C yr^{-1}. On the other hand, according to Seiler and Conrad (1987) the figures are 260 Tg C yr^{-1} and 385 Tg C yr^{-1}, respectively. Logan and her associates also note that the oxidation of man-made nonmethane hydrocarbons creates annually 38.5 Tg CO–C. This means that ~10% of carbon monoxide coming from NMHC is of human origin in an excellent agreement with corresponding values in Table 2.8. On the other hand, on the basis of data one can assume that ~70% of CO formed from CH_4 by chemical reactions in the atmosphere originates from man-made methane molecules.

Thus, we can conclude that ~600 Tg CO–C is produced annually by internal atmospheric sources. Taking into account the emissions at the surface (see Table 2.8) the total annual CO input into the atmosphere is ~1000 Tg C.

The removal of carbon monoxide from the air is due to its chemical reaction with OH radicals. This is proved, among other things, by the inverse relationship between the concentrations of the two species (see Figure 3.3). When the free

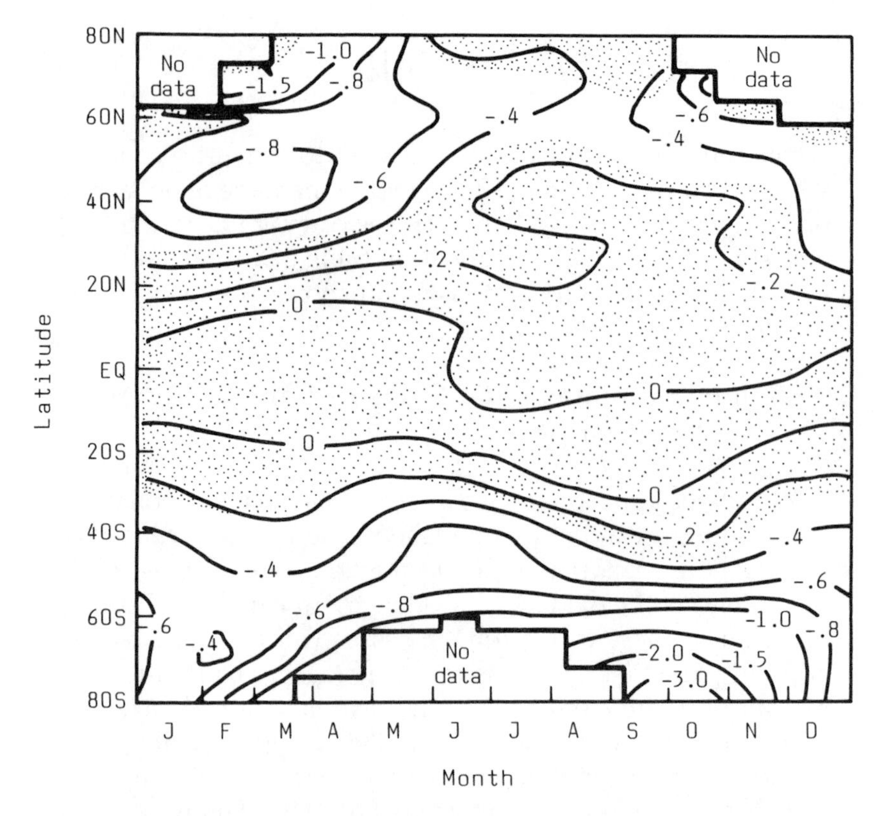

Figure 3.3 Seasonal variation of monthly mean CO mixing ratios observed at Cape Point, South Africa, averaged over the period 1978–1981 (from Seiler et al., 1984). Added is the seasonal variation of OH number density taken from calculations of Logan et al. (1981) and Crutzen (1982).

radical concentration is low (in winter), the CO level* is high and vice versa. The strength of tropospheric internal carbon monoxide sink is estimated to be between 2000 Tg yr^{-1} and 3170 Tg yr^{-1} (Seiler and Conrad, 1987; Logan et al., 1981), which give in carbon equivalents values of 857 to 1359 Tg yr^{-1}. Since the median value of these estimates agrees with the strength of all (surface + internal) sources within a precision of ±15%, we can state that chemical processes remove practically the entire quantity of carbon monoxide from the atmosphere.

* Note that these observations were carried out in the Southern Hemisphere. In the clean troposphere over the Northern Hemisphere the CO concentration is between 150 and 200 ppb (Seiler, 1974).

3.3.4 Nitric Oxide Production During Lightning Discharges

A special class of tropospheric reactions occurs at high temperatures produced during lightning discharges. Under these conditions molecular nitrogen reacts with oxygen molecules in the air to form nitric oxide. It was demonstrated by laboratory studies simulating atmospheric conditions that lightning discharges provide a worldwide source of NO (Chamenides et al., 1977). Legrand and Delmas (1986) assume that lightning at tropical and/or mid-latitudes is the most probable source of nitrate ions found in Antarctic snow.

The global nitrogen fixation can be estimated by using the results of laboratory experiments mentioned as well as the frequency and distribution of lightning discharges all over the world. For their three-dimensional simulation of tropospheric nitrogen compounds Penner et al. (1991) compiled the information available. They accepted a global source strength of 3 Tg yr^{-1} expressed in nitrogen. It goes without saying that this value should be considered with some caution due to uncertainties involved in the estimation.

3.4 FORMATION OF NATURAL AND ANTHROPOGENIC AEROSOL PARTICLES BY CHEMICAL REACTIONS AND CONDENSATION

3.4.1 General Remarks

Photochemical and thermal reactions can produce gases (vapors) in the atmosphere that have low saturation vapor pressure. The vapors condense in the air by bimolecular (water molecules also take part in the process) nucleation* to form minute droplets called "aerosol particles". Generally speaking, the particles formed in this way are very small; their radius is in the range of 10^{-7} to 10^{-6} cm (10^{-3} to 10^{-2} μm). They are termed the *fine* particles to differentiate them from coarse particles of soil origin (see Section 2.5.2). The number of fine

* Nucleation is the initiation of the phase change of a substance during which a more condensed state is formed at a certain point in the less condensed state.

Table 3.1 Number Concentration of Aerosol Particles under Different Conditions[a]

Area of Observations	Concentration (cm^{-3})
Towns	34000
Country air	10000
North Atlantic	600
Oceans in the Southern Hemisphere	400
At 3000 m over continents	300
Pacific Ocean	250
Antarctic summer	200
Antarctic winter	10

[a] Values taken from data compiled by Mészáros (1991a).

particles per unit volume of the air (number concentration) is significant. Their small size and high concentration promote their thermal coagulation, which leads to the formation of larger particles with lower concentration. This coagulation process is due to the random motion from the fluctuating impact of gas molecules on the particles. By coagulation, particles with a radius ~0.1 μm are created. This means that after a certain time the particles are accumulated around this size owing to their collisions.

The number concentration of aerosol particles in the atmosphere is controlled by fine particles, while the mass concentration (particle mass per unit volume of air) is determined by coarse particles. The number concentration is a function of the strength of source and sink mechanisms. Consequently, it varies in space and time in a wide range as Table 3.1 shows. It can be seen that the aerosol concentration is especially high over more polluted continental areas caused by human influences. Observations also indicate (see Mészáros, 1991a) that in summer and during daylight higher concentrations can be measured than in winter and at night. This finding proves the importance of photochemical processes in the formation of tropospheric aerosol particles.

The size of aerosol particles covers several orders of magnitude. For this reason the concentration alone is not sufficient to characterize atmospheric particles. For more complete characterization the size distribution, f(r), must be used:

$$\frac{1}{N}\frac{dN}{dr} = f(r) \qquad (3.34)$$

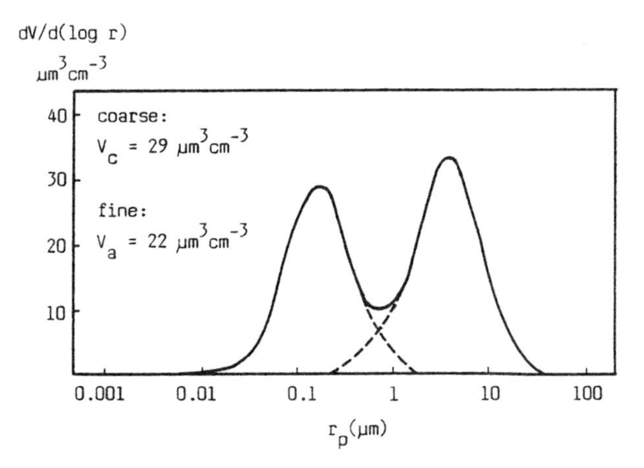

Figure 3.4 Size distribution of the volume (V) of aerosol particles (solid line) in polluted continental air according to Whitby (1978).

where N is the total number concentration and dN is the same parameter for particles with radii between r and r + dr. On the basis of atmospheric observations and laboratory experiments Whitby (1978) proposed that the size distribution of fine particles consists of two log normal distributions. The first distribution is composed of particles formed by nucleation (called the nuclei mode), while the second is due to the coagulation (accumulation mode) in agreement with our previous discussion. The relative significance of these two modes depends on the age of the aerosol.

The number size distribution may be converted to volume or mass size distribution, given an assumed particle shape and density. Since the volume of a spherical particle is proportional to the third power of the radius, the nucleation mode practically disappears in the volume/mass size distribution and the accumulation mode controls the volume/mass of fine particles. Moreover, in a size distribution of this kind coarse particles play an important role, forming a third log normal distribution. Figures 3.4 and 3.5 show the volume size distribution of aerosol particles in moderately polluted continental air (Whitby, 1978) and clean oceanic atmosphere (A. Mészáros and Vissy, 1974). One can see that, with decreasing human influences, the importance of accumulation mode decreases relative to the coarse particle mode.

Aerosol particles control significantly weather and climate. Thus, specific fractions of these particles, the nuclei, can initiate

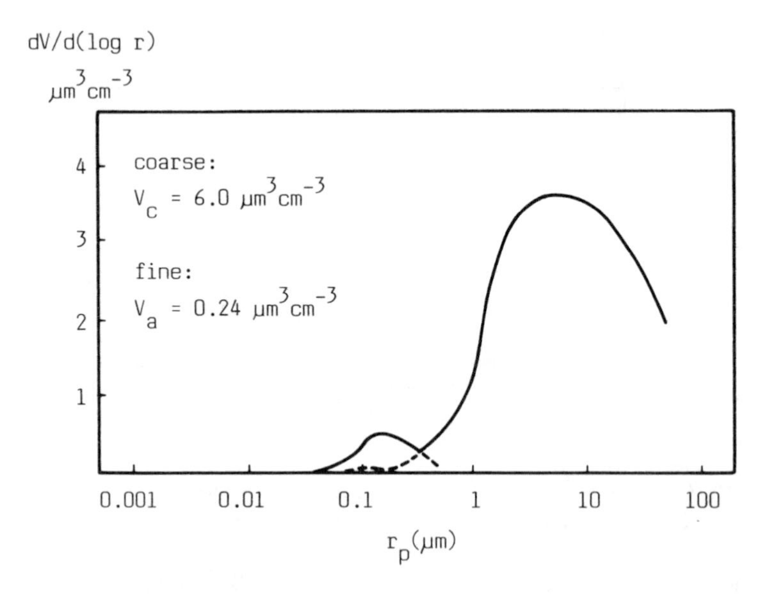

Figure 3.5 Size distribution of the volume (V) of aerosol particles in oceanic air (data of A. Mészáros).

water droplets and ice crystal formation in the air and therefore regulate the radiative properties of the clouds (see Section 6.5.1) as well as the atmospheric water cycle. In addition, particles scatter and absorb solar radiation, influencing the radiative transfer and visibility in the atmosphere. Aerosols are also important from the point of view of atmospheric electricity and radioactivity since gaseous ions and radioactive isotopes in the air are attached to the particles. Obviously human activities modify all these processes and phenomena by releasing into the air precursor gases that are converted chemically into condensable vapors as discussed below.

3.4.2 Formation of Aerosol Particles by Chemical Reactions in the Troposphere

Fine aerosol particles in the air are mostly composed of sulfur and nitrogen compounds, as well as different organics and metals (Finlayson-Pitts and Pitts, 1986; Warneck, 1988; Mészáros, 1991a). Sulfate and nitrate particles are formed from DMS, sulfur dioxide, nitrogen oxides, and ammonia. In these conversion processes hydroxyl and other free radicals play a decisive role.

Under natural conditions, e.g., in oceanic air undisturbed by human activities, sulfur aerosol formation begins with the oxidation of DMS molecules initiated by OH and NO_3 (nitrate) radicals* (Grosjean and Lewis, 1982) as observations indicate (e.g., Berresheim et al., 1991). Sulfur dioxide reacts with OH, which leads to the following reaction sequence (Finlayson-Pitts and Pitts, 1986):

$$OH + SO_2 + M \rightarrow HSO_3 + M \tag{3.35}$$

$$HSO_3 + O_2 \rightarrow HO_2 + SO_3 \tag{3.36}$$

$$SO_3 + H_2O \rightarrow H_2SO_4 \tag{3.37}$$

The end product of these gas phase reactions is sulfuric acid vapor, which immediately condenses due to its very low saturation vapor pressure. The acid droplets formed in this way are neutralized by NH_3 if this gas is available in the atmosphere. Owing to the above process sulfate ions in acidic or neutralized form give the most important fraction of atmospheric aerosol particles in the fine size range.

OH radicals also oxidize nitrogen dioxide to form nitric acid vapor:

$$OH + NO_2 + M \rightarrow HNO_3 + M \tag{3.38}$$

Since the saturation vapor pressure of HNO_3 is higher than that of H_2SO_4, an important part of HNO_3 remains in vapor phase (Mirabel and Jaecker-Voirol, 1988). Its condensation is promoted by existing aerosol particles which serve as nuclei for the phase transition. If the condensation takes place on soil or sea salt particles, HNO_3 in aerosol phase is neutralized by the cations present; in this case nitrate ions are detected in the coarse particle size range. In contrast, particulate nitric acid is found in the fine size range if homogeneous condensation occurs. HNO_3 in fine size range is neutralized by ammonia. The stability of ammonium nitrate particles is low, and HNO_3 and NH_3 can be revolatized as a function of environmental

* Formed by the reaction of NO_2 with ozone: $O_3 + NO_2 \rightarrow NO_3 + O_2$ (see Finlayson-Pitts and Pitts, 1986).

conditions (Stelson et al., 1979). Thus, NH_4NO_3 can form during a cold night and decompose during the day.

There is a considerable body of evidence suggesting that small organic particles are also formed by gas-to-particle conversion (Duce, 1978). Under unpolluted conditions this particle formation is due to the release of natural hydrocarbons from vegetation. In agreement with the original idea of Went (1966), Lopez et al. (1984) assumed that different pinenes emitted by pine forests play an important part in the process. Further, natural forest, brush, and grass fires also provide an important atmospheric aerosol particle source (see Section 2.2.6). In urban and industrial environments the cooling of vapors with low saturation pressure, released during combustion, produces a large quantity of aerosol particles composed mainly of carbonaceous materials. These processes are essential, in particular, since on the surface of elemental carbon (soot) particles formed by condensation, absorbed SO_2 molecules can be converted to sulfuric acid as discussed by Novakov (1984).

A special case of the production of particulate matter by gas-to-particle conversion is provided by the irreversible transformation of trace gases in cloud and fog droplets. A good example of this process is the formation of sulfate from gaseous sulfur dioxide absorbed by cloud/fog elements. It is well documented (Penkett et al., 1979) that the oxidation of SO_2 to form sulfate ions proceeds through the action of oxidizing agents like ozone and hydrogen peroxide formed by chemical reactions in the atmosphere (see Logan et al., 1981). Laboratory experiments (summarized by Beilke, 1985) show that at low pH values of cloud water ($pH < 5.5$) occurring under atmospheric conditions, the oxidation of SO_2 by H_2O_2 is much more effective than the transformation due to ozone molecules (for further details see Section 4.2). O_2 may also be important in oxidation processes if catalyzed by active sites on soot or by transition metals (Jacob and Hoffmann, 1983), at least under more polluted conditions. If NH_3 is also absorbed, the sulfuric acid present is transformed into ammonium sulfate. When the cloud/fog partially or totally evaporates,* ammonium sulfate becomes airborne as demonstrated

* In the case of precipitation from the cloud, ammonium sulfate is removed from the air (see Section 4.2).

by the atmospheric observations of Hegg et al. (1980). In a recent paper Hegg (1991) proposes that even in the cloudy air sulfuric acid particles are formed by gas phase reaction and condensation due to the high in-cloud OH concentrations. OH radicals are created in the cloud under the influence of radiation flux scattered by cloud droplets.

Concerning human effects on aerosol formation one can note the following. As Table 2.8 shows on a global scale, the anthropogenic fraction of volatile organic compound emission is relatively low. On the other hand, a major part of nitric acid formed by chemical reactions remains in vapor phase as discussed above. Thus, we can conclude that human activities affect, first of all, the concentration of sulfate particles in the troposphere by emitting sulfur dioxide in large quantities (see Table 2.8). The residence time of sulfur dioxide and sulfate particles in the air is only several days due to removal processes (see the next chapter). For this reason human modifications are mainly restricted to the continents and industrialized areas as illustrated by Figure 3.6 (Langner and Rodhe, see IPCC, 1990). Accordingly, the mass concentration of excess (non-sea salt) sulfate ions in particulate matter is $<0.1\ \mu gm^{-3}$ ($1\ \mu g = 10^{-6}$ g) in oceanic and clean tropospheric air (Mészáros, 1991a), while it has a magnitude of $1\ \mu gm^{-3}$ over more populated areas of Europe and North America. This indicates that human modifications in sulfate aerosol content over continental regions are much more important than on a global scale. This conclusion involves that the original natural aerosol background in oceanic air over the Southern Hemisphere can be detected; it consists of excess sulfate particles in the fine particle size range (A. Mészáros and Vissy, 1974). In agreement with our discussion this sulfate originates from DMS emitted by the oceanic biota as discussed in several papers (Nguyen et al., 1983; Bigg et al., 1984; Andreae, 1986).

3.4.3 The Stratospheric Sulfate Layer

In the stratosphere an aerosol layer composed of sulfates can be found around the 20 km altitude level as discovered by Junge et al. (1961). During volcanic eruptions stratospheric sulfate particles are formed from SO_2 (see Section 2.3). In volcanically quiet periods sulfur dioxide is the result of the

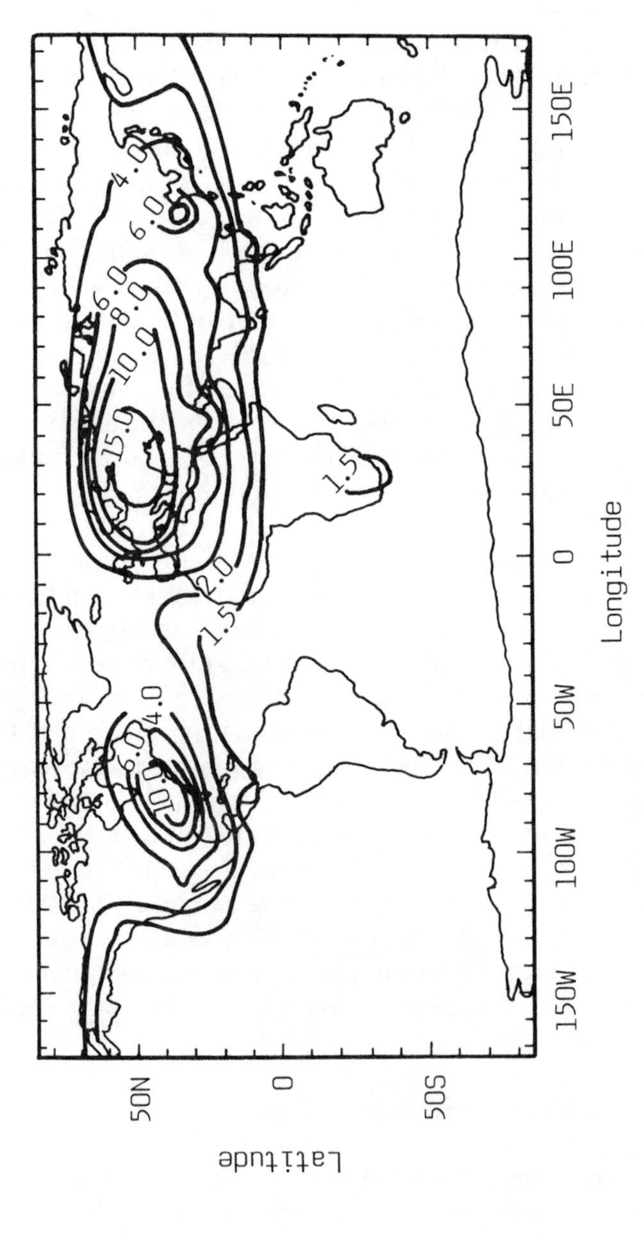

Figure 3.6 Simulated concentration of sulfate at 900 hPa: Ratio of concentration based on total emissions (natural plus anthropogenic) divided by concentrations based on natural emissions in July (calculated by Langer and Rodhe, published in IPCC, 1990).

oxidation of carbonyl sulfide (COS) as first proposed by Crutzen (1976).

COS is emitted in relatively small quantity by oceanic and terrestrial biospheric sources (see the previous chapter). It can also be formed chemically in the air by the oxidation of carbon disulfide (Logan et al., 1979) released from the biosphere.

$$CS_2 + OH \rightarrow SH + COS \tag{3.39}$$

In the lower stratosphere the concentration of COS decreases, while the sulfur dioxide level is practically constant (Warneck, 1988). The sulfur dioxide level is controlled by its formation and chemical removal:

$$COS + h\nu \rightarrow CO + S \tag{3.40}$$

$$S + O_2 \rightarrow SO + O \tag{3.41}$$

$$O + COS \rightarrow SO + CO \tag{3.42}$$

$$SO + O_2 \rightarrow SO_2 + O \tag{3.43}$$

$$SO + O_3 \rightarrow SO_2 + O_2 \tag{3.44}$$

SO_2 molecules formed in this way are converted into H_2SO_4 vapor by Reactions 3.35 and 3.37.

H_2SO_4 is chemically stable, but it can condense to form aqueous sulfuric acid aerosol particles. Theoretical work by Hamill et al. (1977) indicates that heterogeneous condensation of H_2SO_4 and H_2O vapors under stratospheric conditions is much more probable than a homogeneous phase transition. This means that the condensation takes place on existing small aerosol particles. By reviewing the literature Russel and Hamill (1984) conclude that sulfate particles are not neutralized and consist mostly of sulfuric acid. However, there are occasions (Bigg, 1986) when ammoniated aerosols are present due to the injection of tropospheric aerosol particles and ammonia gas into the stratosphere.

The concentration of stratospheric particles is low as compared to aerosol concentrations in the troposphere. However, they play a certain role in the transfer of solar radiation and, as it was recently proposed (see Section 3.2.2), in heterogeneous ozone chemistry. Anthropogenic modifications may occur if human activities increase the emission of carbonyl sulfide and other sulfur gases. In spite of the fact that an increase of nonvolcanic stratospheric sulfate mass (5 ± 2%) has been detected (Hofmann, 1990) between 1980 and 1990, man-made modifications have never been proved until now. Nevertheless, the problem should be kept in mind, since model calculations (Turco, 1982) show that a tenfold increase in atmospheric COS concentration would result in a fourfold increase in the sulfate mixing ratio. Such a change would modify considerably the state of the stratosphere as well as chemical and meteorological processes in this atmospheric domain.

4

Removal of Trace Substances from the Atmosphere

4.1 DRY DEPOSITION

4.1.1 General

Except the escape of some hydrogen (which is balanced by the input of solar protons) after a definite time interval (residence time) atmospheric elements* leave the atmosphere to be deposited onto the Earth's surface. Deposition can occur under dry weather conditions or together with precipitation fall. These two removal types are termed the dry and wet deposition, respectively. More specifically, "*dry deposition* is the aerodynamic exchange of trace gases and aerosols from the air to the surface as well as the gravitational setting of particles" (Hicks et al., 1988). The transfer of gases and fine particles in the direction of the surface is controlled by turbulent diffusion, while gases and particles must be transferred through a thin laminar layer covering all surfaces by, e.g., molecular diffusion or Brownian motion. On the other hand, settling of coarse aerosol particles is the result of the gravitational field.

* Note that chemical reactions in the air provide a sink for compounds but not for elements.

Dry deposition due to turbulent transfer is generally determined by multiplying the surface air concentration by the dry deposition velocity deduced from special micrometeorological measurements made over homogeneous surfaces. Settling velocity of coarse particles can be calculated by balancing the gravitational force with the drag force acting on the particles. Obviously, the settling velocity, V_s, of aerosol particles deduced in this way is a function of their radius, r:

$$V_s = \frac{2}{9}\frac{r^2 \rho_p g}{\mu} \tag{4.1}$$

where g and ρ_p are the gravitational constant and particle density, respectively, and μ is the dynamic gas viscosity (equal to 1.815×10^{-5} N sm^{-2} at a temperature of 20°C).

Dry deposition of gases can be a very effective process if the surface or the ecosystems on the surface remove the molecules considered (e.g., calcareous soils with high alkalinity absorb sulfur dioxide and other acidic species). Further, if the concentration of a compound is high (e.g., in the vicinity of sources) dry deposition is significant and can exceed wet deposition. Generally speaking, dry deposition of gases and particles depends on the state of the atmosphere, the type of underlying surface, and the chemical species. Consequently, its measurement is not an easy task, and dry deposition values must be considered with caution. This means that the precision of global terms discussed below is not better than a factor of 2 or 3.

Finally, we note that by increasing the concentration of different compounds in the air, human activities modify the quantity and quality of atmospheric deposition as well. While deposition is a sink for the atmosphere, it transports materials to aquatic and terrestrial ecosystems altering life processes of plants and animals. The danger caused by man-made modifications is particularly great in the case of acidic substances and toxic metals as will be discussed in Section 6.2.

4.1.2 Trace Gases

Table 4.1 gives the possible magnitude of dry deposition of various trace gases. Values tabulated in the first two lines are taken from the work of Seiler and Conrad (1987). Deposition

Table 4.1 Dry Deposition Velocity (V_d), Concentration (C_A), Dry Deposition Rate (D_d), and Global Dry Deposition (D_{dg}) of Different Atmospheric Trace Gases

Gas	V_d(cm s^{-1})	C_A(μg m^{-3})	D_d(g m^{-2} yr^{-1})	D_{dg}(Tg yr^{-1})
CH_4	1.6×10^{-4}	1.2×10^{3}	5.9×10^{-2}	30 (22.5)
CO	3.0×10^{-2}	8.0×10^{1}	0.76	390 (167)
SO_2	0.8	0.2; 2.0	0.05; 0.5	92 (46)
NO_2	0.25	0.05; 0.5	3.9×10^{-3}; 3.9×10^{-2}	7.1 (2.2)
HNO_3	0.5	0.05; 0.5	7.9×10^{-3}; 7.9×10^{-2}	15 (3.3)
NH_3	1.0	0.01; 0.1	3.1×10^{-3}; 3.1×10^{-2}	5.7 (4.7)

Note: If two concentration values are given, the first refers to oceanic air and the second to continental atmosphere. D_d is calculated by the formula: $D_d = V_d \times C_A$.

[a] The values in parentheses are expressed in C, S, and N.

velocities of SO_2 and NH_3 are based on information published by Beilke and Gravenhorst (1987), while those for nitrogen dioxide and nitric acid vapor are typical values chosen by Penner et al. (1991). Finally, concentration data for gases other than CH_4 and CO are estimated by using the compilation of Mészáros (1992).

Bearing in mind that the values in Table 4.1 are tentative, one can see that gases determining the acidity of the deposition like SO_2, NO_2, HNO_3, and NH_3 are deposited rather quickly from the air, relative to methane and carbon monoxide. As we have seen, the major part of these latter two species are removed by chemical sinks. Considering the table it should be noted that the vertical flux of ammonia is bidirectional; soils and ocean water may act as sources and sinks depending on the characteristics of the air and underlying surface.

4.1.3 Aerosol Particles

Figure 4.1 represents the dry deposition velocities of aerosol particles measured in wind tunnel experiments over water and grass surfaces by Chamberlain (see Beilke and Gravenhorst, 1987). In the figure the gravitational settling velocity calculated by Equation 4.1 is also plotted. It can be seen that with decreasing particle size the deviation between actual deposition and settling is more and more important. In agreement with our previous discussion, it is believed that this deviation is caused by deposition due to turbulent diffusion. Wind tunnel studies

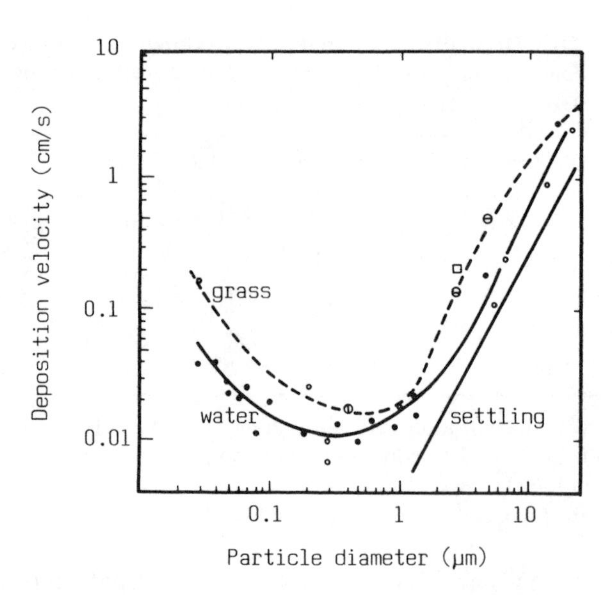

Figure 4.1 Deposition velocities of aerosol particles measured by Chamberlain (see Beilke and Gravenhorst, 1987) in a wind tunnel over smooth surface (water and short grass). The curve settling refers to spherical particles of unit density.

also indicate that the dry deposition velocity increases with decreasing size in the range where Brownian motion becomes important.

We have seen previously that fine tropospheric particles consist mainly of ammonium sulfate (sulfuric acid) and nitrate. Considering that the maximum of the mass of fine particles is in the size range of 0.1 to 0.5 μm (see Figure 3.4) it seems reasonable to suppose on the basis of Figure 4.1 that their dry deposition velocity cannot be greater than 0.1 cm s^{-1}. This is in good agreement with the estimate of Garland (1978) and the observation of Davies and Nicholson (1982), giving an average value of 0.08 cm s^{-1} for sulfate particles. If, on the basis of Figure 4.1, we assume an average of 0.03 cm s^{-1} and take into account appropriate atmospheric concentrations for nonpolluted tropospheric air (see Mészáros, 1991a), we can state that the dry deposition of fine particles can be neglected on a larger scale compared to that of precursor gases. Low dry depositions also mean that the vertical flux of fine particles is controlled by wet removal processes as we will see in the next subsections. Before a final conclusion, however, it should be considered that some field observations resulted in higher dry deposition rates than the above value as compiled by Voldner et al. (1986). The reason for this disagreement is not clear and its explanation needs further efforts.

On the other hand, it is generally accepted that the dry deposition of coarse particles is significant. The deposition of dust particles originating from Asian deserts and the Sahara can be significant under suitable weather conditions, even rather far from the source regions (see Section 2.5.2). According to a recent study of Schneider et al. (1990) carried out in Central North Pacific, dust particles in the size range 2 to 4, 4 to 8, and 8 to 15 μm have a deposition velocity of 0.3, 1.2, and 2.6 cm s^{-1}, respectively. Since human activities in semiarid areas (e.g., burning of vegetation, grazing) lead to desertification, it cannot be excluded that the windblown dust burden in the air, and consequently the dust deposition, will be increased in the future.

4.2 WET SCAVENGING OF TRACE SUBSTANCES

4.2.1 In-Cloud Scavenging of Aerosol Particles

Wet removal of trace substances takes place partly in clouds and partly below the clouds. In the first case, called the "in-cloud scavenging", aerosol particles and trace gases are removed from the air by cloud droplets and ice crystals, while in the second trace materials are scavenged by falling precipitation elements (rain drops and snow flakes). This latter type of wet removal is termed the sub-cloud scavenging.

The removal of aerosol particles begins together with cloud formation. Cloud droplets form on special aerosol particles, called the *cloud condensation nuclei* (CCN). Generally speaking, the nuclei consist of water soluble particles, mostly sulfates, since soluble nuclei are more active in condensation than insoluble particles. This means that droplets come into being at lower supersaturations on soluble particles than on insoluble ones. In other words soluble CCN have lower critical supersaturation. Particle size is also an essential parameter, owing to the fact that larger particles have higher activity in cloud formation than smaller nuclei.

Cloud formation is the result of the updraft of humid air, which is warmer than its surroundings. Because of cooling, the air becomes supersaturated with respect to water vapor. The supersaturation first increases then decreases due to water vapor consumption by nuclei. Only those particles whose critical supersaturation is equal to or less than the maximum supersaturation reached can serve as CCN. If the number of such CCN is high the water vapor quantity available for condensation is distributed on many nuclei. Consequently, the cloud is composed of small droplets with high concentration. Since the probability of precipitation formation by collision of cloud droplets is directly proportional to their size, the precipitation formation ability of such clouds is low and their lifetime is long relative to clouds consisting of larger droplets of low concentration.

The number of active nuclei depends on the properties of the particles and on the updraft velocity. However, one can postulate that soluble particles having a dry radius* larger than ~0.01 to 0.05 μm are activated under normal conditions. Since the major part of the mass of fine particles is above this size, we can conclude that soluble particles are removed very effectively by condensation. It goes without saying that coarse soluble nuclei like sea salt particles serve always as CCN, but they give only a minor fraction of CCN due to their low number concentration relative to the number of cloud droplets in clouds. However, sea salt particles can play a certain role in the formation of larger cloud droplets important for rain formation (the details of the above discussion can be found in Mészáros, 1991b).

The presence of ice crystals in clouds may also initiate rain formation. Ice crystals have a lower saturation level than supercooled liquid droplets. Consequently, the crystals absorb water vapor molecules, which lowers the vapor pressure in their surroundings. In this environment the droplets evaporate while the crystals grow. Ice crystals form on special aerosol particles called ice nuclei. The activity of particles in ice formation and their mode of action depend on many factors, but it is well documented that their number increases with decreasing temperature (see Vali, 1991).

---

* The actual radius of soluble particles depends on the relative humidity of the air, since they occur in liquid phase even below the saturation level.

This short discussion shows that cloud droplets and ice crystals contain foreign materials even at the beginning of cloud formation; however, still numerous particles remain in the cloudy air (interstatial particles). Very small particles, nonactivated in condensation, have an important Brownian motion and may have a relatively high number of concentrations depending on the age of the aerosol. For these reasons they coagulate with cloud droplets and ice crystals, and their number decreases exponentially with time. Thus, due to coagulation they are also removed from the air and are imbedded in cloud elements. Finally, an important part of aerosol particles is transferred from the air to cloud water. The fraction removed is significant, in particular, in clean air under remote conditions.

4.2.2 In-Cloud Scavenging of Trace Gases

After cloud formation the sorption of soluble trace gases begins immediately. Under ideal conditions the concentration in liquid phase can be calculated by Henry's law, which shows that the concentration of the gas in liquid phase is directly proportional to the partial pressure in the gas phase. Gas scavenging is particularly important in those cases in which an absorbed gas reacts irreversibly in the liquid phase with another constituent. Example of this type of transformation is provided by the conversion of sulfur dioxide into sulfuric acid (sulfate ions). Since all the processes taking place in clouds depend on the concentration of hydrogen ions, the determination of pH (the negative exponent of the hydrogen ion concentration) is essential to evaluate the efficiency of gas scavenging.

The pH of cloud water is first of all controlled by the absorption of carbon dioxide molecules, which yields by dissociation bicarbonate and carbonate ions. The hydrogen ion concentration (bracket denotes molar concentration: M = mole per liter) in equilibrium is

$$\left[H^+\right]=\left[OH^-\right]+\left[HCO_3^-\right]+2\left[CO_2^{2-}\right] \tag{4.2}$$

By using appropriate physicochemical data and atmospheric CO_2 concentrations, the equilibrium $[H^+]$ can be calculated. The result is a pH value of 5.6 at a temperature of 10°C. This

Table 4.2 Numerical Values of Rate Constants in Reactions 4.3 and 4.4 at 25°C[a]

Rate Constants	Numerical Values[b]
k_a	2.0×10^4
k_b	3.2×10^5
k_c	1.0×10^9
k_d	5.6×10^9

[a] Data taken from Warneck, 1988.
[b] Expressed in L mol^{-1} or L^2 mol^{-2} where applicable.

pH value is considered as the atmospheric neutral point. This means that atmospheric waters (clouds, fogs, precipitation) are acidic if their pH is lower than the above value.

When sulfur dioxide is absorbed by cloud droplets, bisulfite (HSO_3^-) and sulfite (SO_3^{2-}) ions are formed from aqueous SO_2 molecules (SO_{2aq}). Many research efforts have been done to determine the details of the oxidation of these species. It is now well documented (e.g., Penkett et al., 1979) that their oxidation to form sulfate ions proceeds through the action of oxidizing agents like ozone and hydrogen peroxide* (also absorbed by the liquid phase). Their loss is given by the following rate expressions (Warneck, 1988)

$$
\begin{aligned}
-\frac{d[S(IV)]}{dt} &= k\left[O_{3aq}\right][S(IV)] \\
&= \left[O_{3aq}\right]\left[k_a\left(SO_{2aq}\right)+k_b\left(HSO_3^-\right)+k_c\left(SO_3^{2-}\right)\right]
\end{aligned}
\tag{4.3}
$$

$$
-\frac{d[S(IV)]}{dt} = k'\left[H_2O_2\right][S(IV)] = \frac{k_d\left[HSO_3^-\right]\left[H_2O_2\right]\left[H^+\right]}{0.1+\left[H^+\right]} \tag{4.4}
$$

where S(IV) denotes sulfur of four valances (SO_2, bisulfite, and sulfite). The rate constants of these processes are given in Table 4.2. It follows from these reactions and rate constants tabulated

* Hydrogen peroxide is formed in the air by the reaction $HO_2 + HO_2 \rightarrow H_2O_2 + O_2$ (e.g., Logan et al., 1981).

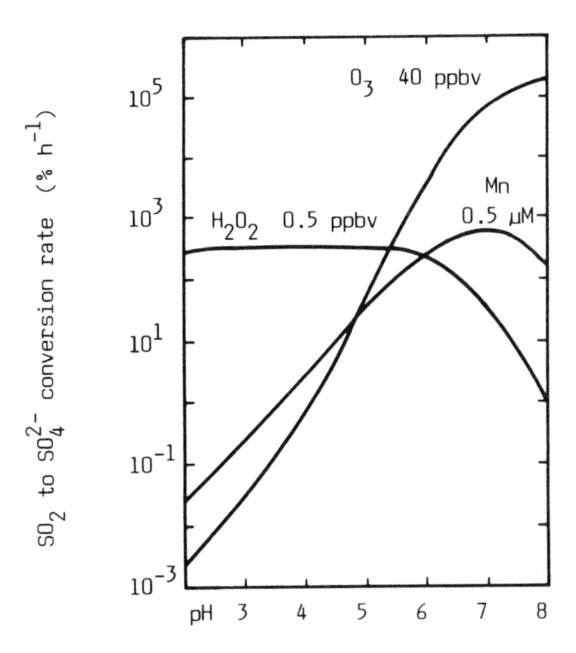

Figure 4.2 Aqueous oxidation of SO_2 in clouds as a function of pH; relative rates for reactions of dissolved SO_2 with ozone, H_2O_2 and oxygen, the latter catalyzed by manganese for the concentrations indicated (Warneck, 1988).

that ozone oxidizes first of all sulfite ions, while hydrogen peroxide reacts solely with bisulfite. The process (Reaction 4.3) is more rapid at high pH, when sulfite formation is important. On the other hand, the rate of Reaction 4.4 is directly proportional to $[H^+]$ for $[H^+] < 10^{-2}$(pH > 2). This means that at high pH values Reaction 4.3 is dominant, while for pH below 5.4 the process (Reaction 4.4) controls the formation of sulfate ions (see Figure 4.2). Considering the fact that the pH of cloud water is generally below 5.4, it can be concluded that hydrogen peroxide plays an important part in the oxidation of SO_2 in clouds.

It can be noted that the catalyzed oxidation of S(IV) in cloud water is also possible. Such a catalyst can be provided by iron or manganese. As Figure 4.2 shows under atmospheric conditions (pH below 5.6) the rate of this reaction depends strongly on pH. Its effects can be generally neglected, except on local scale when the air is polluted by the catalytic elements in question.

While the oxidation of NO_2 absorbed by cloud water is theoretically possible, it can be demonstrated that NO_2 is transformed more effectively to nitric acid in gaseous phase as discussed in Section 3.3.1. This implies that the nitrate content of cloud water originates mainly from the removal of HNO_3 vapor (e.g., Chang, 1984).

Parallel with the removal of SO_2 and HNO_3, ammonia gas is also absorbed by cloud water. Ammonia, the only gaseous alkaline species, dissociates in aqueous phase to give ammonium ions:

$$NH_3 \cdot H_2O \rightarrow NH_4^+ + OH^- \tag{4.5}$$

This results in the neutralization of sulfuric and nitric acid. It is believed that this process removes a significant part of NH_3 molecules from the atmosphere.

4.2.3 Sub-Cloud Scavenging

There are two possibilities concerning the future fate of clouds. One possibility is that the cloud evaporates and absorbed trace constituents again become airborne. However, a new aerosol is produced in this way compared to the size distribution before cloud formation, since one drop generally captures several aerosol particles and reactive trace gases (SO_2, NH_3, HNO_3) are transformed in cloud water. At the same time after dissipation the concentration of reactive gases is lower than before cloud formation owing to irreversible removal processes.

On the other hand, if the cloud precipitates, the materials absorbed are carried out by precipitation to the surface of the Earth; they are definitively removed from the air. We have to emphasize that precipitation elements further scavenge trace materials from the air between the cloud base and the surface. The quantity of substances scavenged in this way is added to the material amount removed in clouds. If the air is not saturated below the clouds, which is generally the case at the beginning of rainfall, the partial evaporation of drops increases the concentrations. Thus, the concentration of different substances in precipitation collected at the surface is the result of three processes: in-cloud and sub-cloud scavenging as well as drop evaporation.

Aerosol particles below the clouds are captured by precipitation elements owing to the difference between falling speeds of the aerosol particles and the raindrops or snow crystals. More precisely, precipitation elements overtake the particles, which are impacted against the drops due to their inertia; the higher the inertia (size) of the particle the higher the probability of such a collision. For this reason mostly coarse particles are removed by sub-cloud scavenging. In other words, the concentration of sea salt particles (oceanic conditions) and particles of soil origin (continental conditions) is efficiently lowered by the process. Since these particles are generally alkaline, sub-cloud scavenging increases the pH of precipitation under normal conditions.

The absorption of trace gases below the cloud base is controlled by the same processes as in-cloud scavenging. The sub-cloud scavenging of trace gases may be important, in particular, if the concentration of the gas considered increases with decreasing height—that is in more polluted air—where sources can be found at the surface.

4.2.4 Wet Deposition Modeling

If we want to describe theoretically the wet deposition, D_w, as a whole, we assume that in an air layer with a depth of h it depends only on two parameters: the concentration in the air, C_A, of the substance considered and the rainfall rate, R (see, e.g., the recent review of Smith, 1991):

$$D_w = C_L R = \omega C_A R \tag{4.6}$$

where ω is the so-called washout ratio defined as the ratio of the concentration in liquid water, C_L, and in the air of the species considered. By analogy to the dry deposition velocity

$$\frac{dD_w}{dt} = v_w C_A = \omega C_A \frac{dR}{dt} \tag{4.7}$$

Thus, for characterizing the dynamics of the wet deposition it is crucial to determine the value of the washout ratio. Without further details we note that its magnitude is around 10^5 if C_A and C_L are expressed in g m^{-3}. If the washout ratio is known from independent atmospheric observations or laboratory experiments, by means of Equation 4.7, we can simulate in a simple way the variation of the wet deposition.

4.2.5 The Magnitude of Wet Removal: Precipitation Chemistry Measurements

The chemical composition of precipitation water gives information on the self-cleansing rate of the atmosphere. Additionally, precipitation chemistry observations provide data on the material quantity received by different ecosystems from the atmosphere. For these reasons many national and international research and monitoring programs have been initiated to measure the chemical composition of precipitation (see Mészáros, 1992). Special attention has been devoted to the measurement and interpretation of hydrogen ion concentration as we will see in Section 6.2.

When measuring the composition of precipitation, the concentrations of different ions (sulfate, nitrate, chloride, ammonium, sodium, and soil derived species) in unit volume of the water is determined. If it is multiplied by the amount of precipitation fallen during the sampling time, the *wet deposition* is obtained (see Equation 4.6). The unit of wet deposition ($g\ m^{-2}\ yr^{-1}$) may be similar to the rate of dry deposition used in Table 4.1.

The chemical composition of precipitation water is relatively well-known for Europe and North America (Nodop, 1986; Barrie and Hales, 1984). As an illustration, typical data obtained in Southern Sweden and New Hampshire are presented in Table 4.3. For the evaluation of anion concentrations we note that sulfate and nitrate ions come mostly from anthropogenic aerosol particles and precursor gases, while on a larger scale the major part of chloride ions originates from sea salt containing sodium chloride. If we suppose that sodium is entirely of maritime origin, from the data tabulated for Sweden and New Hampshire a continental chloride concentration of 3 μ eq L^{-1} and 9 μ eq L^{-1} is calculated, respectively. This means that some chloride of continental (anthropogenic?) origin can also be detected in precipitation. Among cations, ammonium is a biogenic product; however, its biological sources are modified significantly by man, as we have previously seen. Other cations, except sodium, are obviously of soil origin. Data tabulated make it evident that the high concentration of

Table 4.3 Average Concentration of Precipitation Water Collected in Southern Sweden[a] and in New Hampshire[b]

Ions	Sweden	New Hampshire
H^+	52 (pH = 4.28)	74 (p = 4.13)
Ca^{2+}	14	8
Mg^{2+}	8	4
K^+	3	2
Na^+	15	5
NH_4^+	31	12
Sum of cations	123	105
SO_4^{2-}	70	60
NO_3^-	31	23
Cl^-	18	14
Sum of anions	119	97

Note: The values are expressed in μeq L^{-1}. For univalent ions 1 eq is equal to 1 mol while for bivalent ions (calcium, magnesium, and sulfate) the number of moles must be multiplied by two to obtain the number of equivalents.

[a] Swedish Ministry of Agriculture, 1982.

[b] Likens et al., 1977.

hydrogen ions (low pH) is due to sulfuric and nitric acid. The sum of anions and cations are nearly equal at both places, indicating that all essential components were analyzed. This means implicitly that the role of other species, like organic acids important under tropical conditions (Galloway et al., 1982), is insignificant over Europe and North America.

In contrast to Europe and North America the distribution of wet deposition on a worldwide scale is not well established. However, in spite of the small data base, efforts were made to determine the global distributions of sulfur and nitrogen containing ions due to the importance of the question. Figures 4.3 through 4.5 show the results obtained by Georgii (1982) and Böttger et al. (1978). Although these distribution patterns should be considered with caution, on the basis of Figure 4.3 we can assume that the oceanic wet deposition of sulfate (expressed in sulfur) is around 0.05 g S m^{-2} yr^{-1}, while the corresponding figure for the continents is 0.3 g S m^{-2} yr^{-1}. This difference is obviously explained by the distribution of anthropogenic sources on the Earth's surface. Further, according to Figure 4.4 one can estimate that the oceanic and continental wet deposition rates of ammonium-nitrogen are equal to 0.02 and 0.2 g N m^{-2} yr^{-1}, respectively. Finally, data in Figure 4.5 make it possible

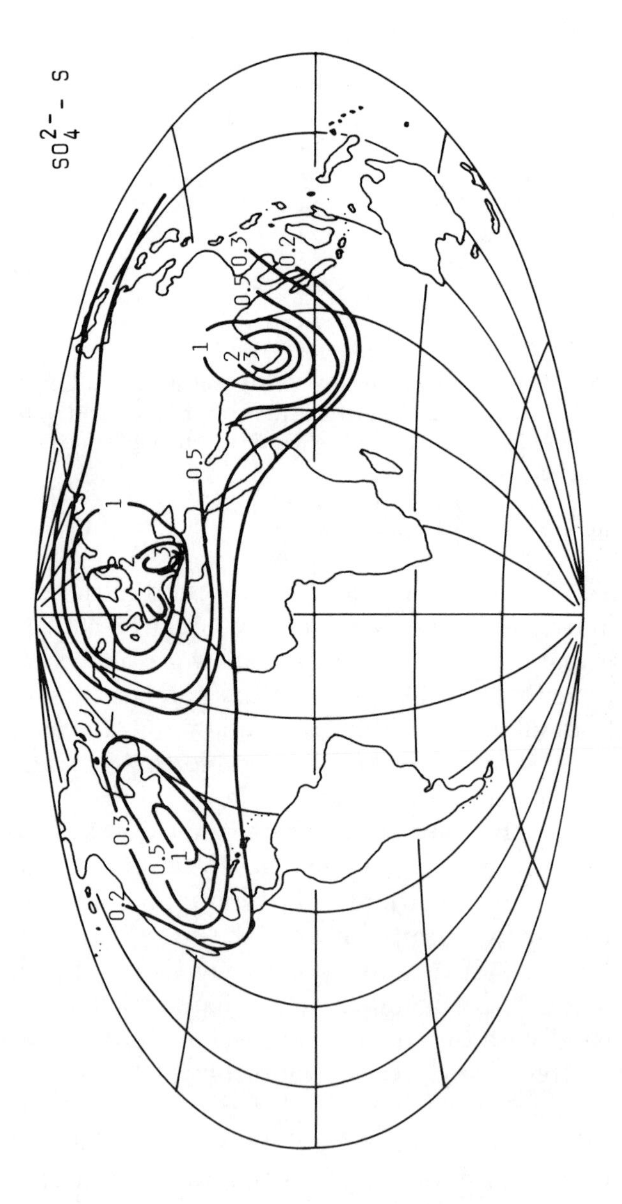

Figure 4.3 Sulfate in rain water. Deposition in units of g S m^{-2} yr^{-1} (from Georgii, 1982).

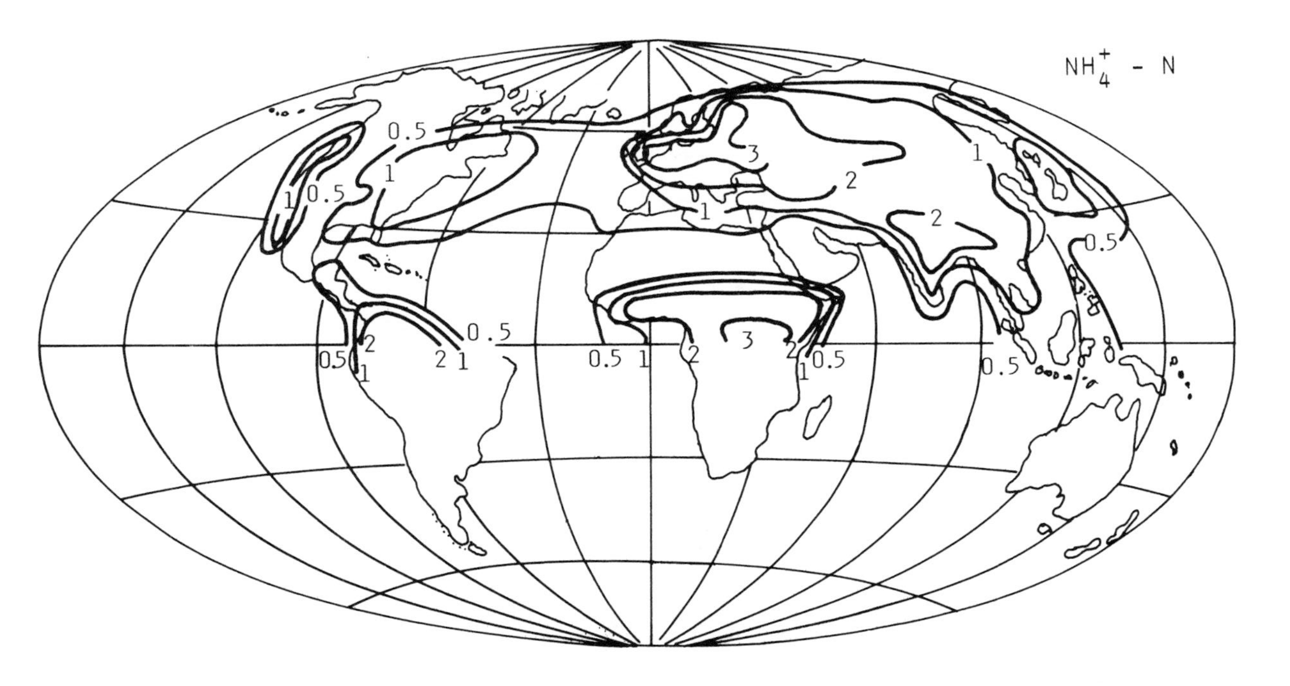

Figure 4.4 Global distribution of wet NH_4^+ deposition according to Böttger et al. (1978), derived from measurements during the period 1950–1977. The deposition rate in units of 100 mg m^{-2} yr^{-1}.

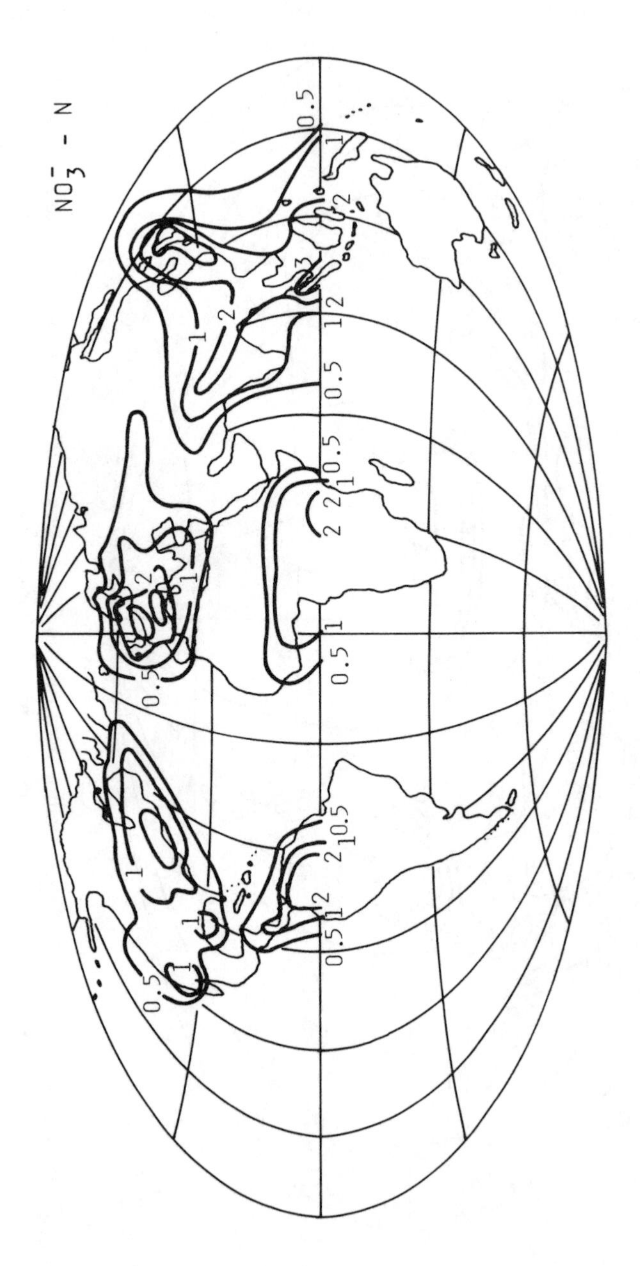

Figure 4.5 Global distribution of the wet deposition rate of NO_3^- nitrogen for the Nothern Hemisphere in units of 100 mg m^{-2} yr^{-1}, derived by Böttger et al. (1978) from measurements during the period 1955–1977.

Table 4.4 Wet Deposition Rate (D_w) and Global Wet Deposition (D_{wg}) of Sulfur and Nitrogen

Species	$D_{w\ ocean}$ ($g\ m^{-2}\ yr^{-1}$)	$D_{w\ cont.}$ ($g\ m^{-2}\ yr^{-1}$)	D_{wg} ($Tg\ yr^{-1}$)
SO_4^{2-}-S	0.05 (3.1)[a]	0.3 (19)	62
NH_4^+-N	0.02 (1.4)	0.2 (14)	37
NO_3^--N	0.02 (1.4)	0.2 (14)	37

[a] Data in parentheses expressed in milliequivalents per square meter per year.

that the nitrate-nitrogen wet deposition rates are similar to those proposed for ammonium-nitrogen.

Table 4.4 summarizes this tentative information on wet deposition rates together with the global values calculated. Comparing these values with data tabulated in Table 4.1 one can conclude that wet deposition is the main removal process of sulfur and nitrogen compounds on a global scale. While direct estimations are not available, it is obvious that the anthropogenic fractions of these deposition terms is identical with those calculated for the emissions (see Table 2.8).

As it was mentioned in Section 2.1 the analyses of snow samples from Greenland indicate that the concentration of sulfate and nitrate ions in wet deposition has increased in the Northern Hemisphere since the end of the last century (Nefter et al., 1985). This is due to the intensification of the strength of anthropogenic emission of SO_2 and NO_x. A second reason may be the increase in the concentration of H_2O_2 in clouds and precipitation, also demonstrated by the analyses of ice core samples collected in Greenland (Sigg and Nefter, 1991). Higher hydrogen peroxide concentration has led to higher oxidation capacity of atmospheric waters as discussed in the previous subsection.

Since the time period of precipitation chemistry measurements, several studies have been made to look for possible trends in data. Thus, it was found (Brimblecombe and Pitman, 1980) that in England nitrate concentration has increased from the end of the last century until now, mainly in spring months. A similar trend was reported on the basis of the results of Hungarian observations (Horváth, 1983). Hungarian data also reveal that during this century the ammonium level of precipitation has remained essentially constant. Precipitation chemistry measurements carried out in New Hampshire since 1964 indicate some increase in nitrate concentration and a relatively constant pH level (Stensland et al., 1986). At the same time the concentration of sulfate ions has decreased significantly (2% per year) in accordance with the trend of SO_2 emissions in the northeastern part of the U.S. (Husar, 1986). According to studies made in Scandinavia the sulfate concentration in precipitation increased by 50% during the late 1950s and 60s, while it declined by 20% since the early 1970s in agreement with changes in SO_2 emissions (Rodhe and Granat, 1984). All these data demonstrate anthropogenic modifications of sulfur and nitrogen wet deposition. Among other things, these modifications resulted in the acidification of the environment on regional and continental scales.

Beside the main components discussed above the study of the deposition of carbonaceous materials and different metals is also important in solving environmental problems. The deposition rate of elemental carbon, occurring only in the aerosol phase, was determined by Ogren et al. (1984). They found that in Seattle, Washington and at a rural site in Sweden the wet deposition flux of elemental carbon is equal to 0.05 g m^{-2} yr^{-1}. The deposition rates of atmospheric trace metals measured under different conditions are evaluated by Galloway et al. (1982). According to their compilation at remote places, the bulk (dry + wet) deposition of, e.g., lead and copper, is around 10^{-4} g m^{-2} yr^{-1}. On the other hand, Duce (1986) proposes a mean Al deposition rates with a magnitude of 10^{-2} g m^{-2} yr^{-1} for the air above the tropical North Pacific, while the corresponding value he gives for the tropical North Atlantic is about four times greater.

5

Atmospheric Cycles and Their Changes

5.1 THEORETICAL CONSIDERATIONS

As we have seen, different constituents are added continuously to atmospheric reservoir[1] (sources[2]). At the same time, they are unceasingly removed from the air by sink processes. Thus, the time, t, variation of the mass, M, of a given species is determined by the difference of the intensity of sources, S, and sinks, R. However, if we consider atmospheric domains or any part of the atmosphere (e.g., continental troposphere) separately, the material transport should also be taken into account:

$$\frac{dM}{dt} = A_{in} - A_{out} + S - R \qquad (5.1)$$

where A_{in} and A_{out} are the mass entering or leaving the compartment by transport (advection). In equilibrium ($dM/dt = 0$) the knowledge of the strength of sources and sinks makes it possible to determine the difference of advection terms.

If we are interested in the transport of a material in more detail, the process should be characterized by the continuity equation, which gives the concentration variation as a function of regular (wind) and irregular (turbulent diffusion) motions in the air. By combining the transport equation with kinetic equations (giving the chemical transformations), as well as with wet and dry removal dynamics, a model can be constructed that makes it possible to calculate the concentration and deposition distribution if the emission field is known. The integration of model equations give the total mass and total deposition of the substance considered. If only vertical variations are taken into account and the surface air is characterized by global averages of meteorological and chemical variables the model is of one-dimensional. In a two-dimensional model latitudinal variations are also considered, while a three-dimensional model describes the atmosphere as a whole.

Such numerical models are very useful for studying the atmospheric circulation of different constituents. Models also provide a good tool to estimate anthropogenic modifications; however, the results obtained in this way should be carefully verified by atmospheric observations.*

As it was mentioned previously several times the pathway and effects of different trace substances are controlled in a great measure by their atmospheric residence time. This important parameter can be determined by dividing the total mass of the species in the atmosphere (or in a part of the atmospheric reservoir) by the formation or removal rates. Thus, in equilibrium the residence time, τ, is calculated by the following simple equation:

$$\tau = \frac{M}{S} = \frac{M}{F} \tag{5.2}$$

In summary one can say that *budget calculations* provide "a systematic comparison among sources, sinks and burdens of a particular compound or element in a portion of the atmosphere... The appropriate horizontal scale depends on the residence time which determine the characteristic transport distances" (Galloway and Rodhe, 1991). The aim of this chapter is to present the tropospheric cycles of the most important carbon, nitrogen, and sulfur compounds on the basis of our

---

* There is no intention here to discuss budget modeling in further details. The interested reader is referred to Pearman et al. (1983), Langner and Rodhe (1991), Penner et al. (1991), and to their references.

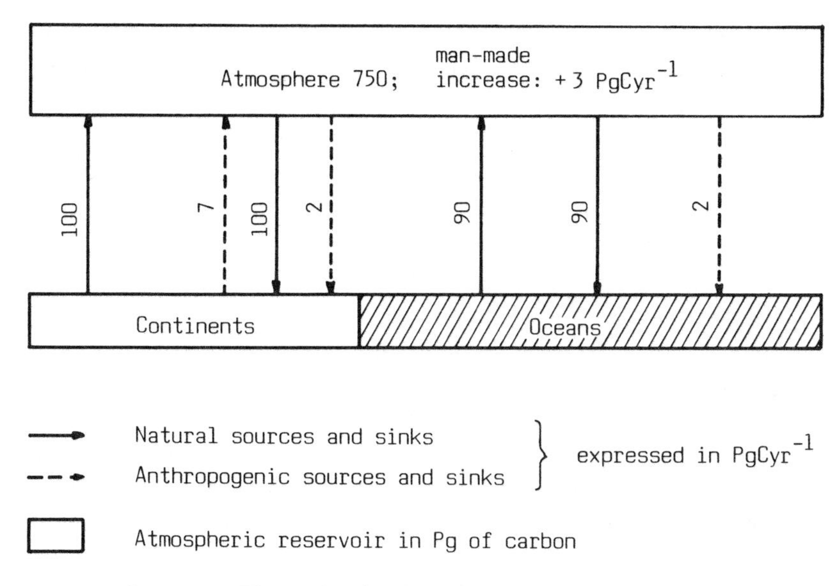

Figure 5.1 The cycle of carbon dioxide in the atmosphere.

previous discussions. In the following discussion global budgets are calculated. Considering the residence time of different species, the global scale is appropriate mainly for carbon compounds. Nevertheless, this approach also is capable of estimating overall worldwide modifications in sulfur and nitrogen cycles. However, some results of regional budgeting of sulfur and nitrogen species will be mentioned. For further details of regional budgets the reading of the review paper of Galloway and Rodhe (1991) is proposed.

5.2 THE CARBON CYCLE

5.2.1 Carbon Dioxide

Figure 5.1 shows the atmospheric cycle of carbon dioxide in a simplified form. The information used to construct the figure is taken from IPCC (1990). One can see that without man-made effects both the terrestrial and oceanic biogenic sources are balanced by terrestrial and oceanic sinks, respectively. Data also indicate that the anthropogenic fraction of the emissions is small. Nevertheless it has disturbed the natural equilibrium. It follows from the figure that carbon dioxide produced by human activities is absorbed by the continental biosphere and the oceans in equal amounts. Nearly half of man-made CO_2

remains in the atmosphere, increasing the CO_2 level in the air (see Section 2.1). This amount is well determined by atmospheric observations.

By applying Equation 5.2 an atmospheric residence time of 4 years is formally obtained. In contrast to this, in Table 2.1 values between 50 and 200 years can be found. This is due to the fact that these latter values give the time during which the concentration of CO_2 would be adjusted to equilibrium if sources and sinks were changed (IPCC, 1990). These times were calculated by means of suitable numerical models. For the sake of simplicity the oceanic part of the carbon cycle in nature is not represented in Figure 5.1. We note, however, that the CO_2 transport by water motions in the oceans play a crucial part in the control of atmospheric carbon dioxide burden. This is explained by the fact that the CO_2 uptake by the oceans, aside from weather conditions, depends on the difference between atmospheric partial pressure and equilibrium partial pressure in water. Since, the latter is controlled by the exchange between surface and deeper water, dynamic processes in the oceans are very important from the point of view of CO_2 absorption from the atmosphere. The role of the marine biota in this exchange is also essential, since in surface water biological processes determine the organic carbon formation from CO_2 absorbed from the air and the sedimentation rate of organic decay products to deeper water. Thus, it can be stated that the fate of anthropogenic carbon dioxide in the atmosphere depends in a large measure on oceanic processes.

The model calculations of Tans et al. (1990) indicate that theoretical CO_2 distributions are consistent with observed north-south atmospheric concentration gradients only if sinks in the Northern Hemisphere operate more efficiently than in the Southern Hemisphere. Observations also show that the more efficient sink processes cannot be explained by oceanic uptake. For this reason Tans and his associates conclude that a large amount of the carbon dioxide should be absorbed on the continents probably by terrestrial ecosystems. Thus, if we calculate the carbon dioxide amount emitted between 1980 and 1989 by anthropogenic sources (fossil fuel + deforestation) and take into account the accumulation in the atmosphere and uptake by the oceans, a net positive imbalance of 1.6 Pg C yr^{-1} is obtained. A similar calculation carried out by Detwiler and Hall (1988) resulted in a missing amount of 2.8 Pg C yr^{-1}, which

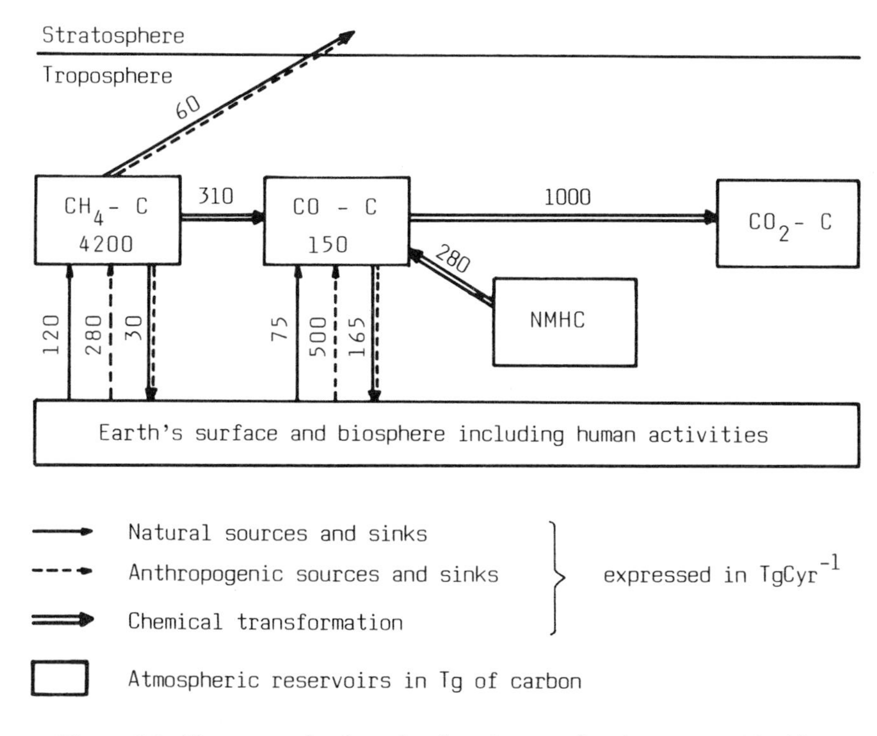

Figure 5.2 The tropospheric cycle of methane and carbon monoxide. Note that the units in Figures 5.1 and 5.2 are different.

is bit higher but not inconsistent with the sink term in Figure 5.1. Obviously, more research has to be done to clarify the details of the fate of this excess CO_2.

5.2.2 Tropospheric Budget of Methane and Carbon Monoxide

In addition to carbon dioxide, methane and carbon monoxide are the most important carbon compounds in the atmosphere. The tropospheric budget of these two species is summarized in Figure 5.2. One can see that these cycles are in relation with the cycle of CO_2. However, the carbon dioxide quantity formed annually from carbon monoxide can be neglected compared to the intensity of other CO_2 sources (see also Figure 5.1).

Concerning Figure 5.2 we note that different source and sink terms are the rounded off values discussed in the previous chapters. The exception is the conversion of CH_4 to CO and of NMHC to CO. These conversions were normalized by requiring the methane and carbon monoxide reservoirs to be balanced.

However, the values obtained in this way are very close to the figures presented on the basis of more direct arguments in Sections 3.3.2 and 3.3.3.

Moreover, the tropospheric burden of methane and carbon monoxide was calculated by the method outlined in Section 3.3.2. For this calculation, average methane and carbon monoxide concentrations of 1.7 ppm and 0.08 ppm were used. This latter value, also used for dry deposition calculations (Table 4.1), is based on the results of atmospheric observations made by Seiler (1974). If the quantities in tropospheric reservoirs are divided by the source or sink strengths, residence times of 10.5 and 0.13 yr (one and half month) are received for CH_4 and CO molecules, respectively. Since the tropospheric mixing time is around 1 yr, methane molecules are well mixed in the troposphere and can reach the stratosphere as illustrated in Figure 5.2. In contrast to CH_4 the concentration of carbon monoxide is more variable in space and time (Seiler, 1974) due to its shorter reside time.

Figure 5.2 also shows that a major part of CH_4 molecules is converted to CO. On the other hand, the main sink for CO is its chemical transformation into carbon dioxide molecules absorbed subsequently by the terrestrial biosphere and the oceans. This indicates that the carbon released by the biosphere and oceans to the atmosphere in the form of CH_4 and CO returns into the emitting reservoirs as CO_2. A part of this carbon amount is of anthropogenic origin; however, as it was said, the carbon dioxide quantity formed in this way is small. Thus, the environmental consequences of these transformations on the biosphere are minor. However, methane and carbon monoxide of human origin are dangerous species since they modify the greenhouse effect and the rate of chemical reactions in the atmosphere.

5.3 TROPOSPHERIC PATHWAYS OF NITROGEN AND SULFUR COMPOUNDS

5.3.1 Ammonia and Nitrogen Oxides

The tropospheric cycle of ammonia/ammonium and nitrogen oxides is illustrated in Figure 5.3. In the figure the cycle of nitrous oxide inert in the troposphere is not plotted. It

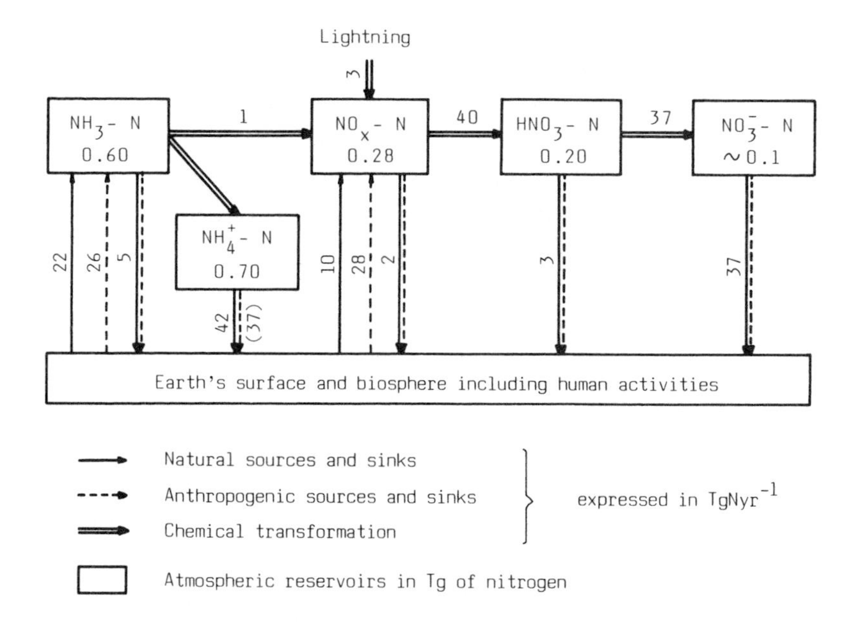

Figure 5.3 Cycle of ammonia/ammonium and nitrogen oxides in the troposphere. For NO_x "x" is equal to 1 or 2; for NH_4^+ deposition, the value given at the left side of the arrow is calculated to balance the budget, while the value in parenthesis is based on direct estimation.

is assumed that NO_x formed in the stratosphere from N_2O is of minor importance for tropospheric nitrogen budget. This assumption is supported by recent model calculations of Penner et al. (1991). The values of emissions and depositions in Figure 5.3 are based on discussions in previous chapters. However, the deposition term given for ammonium is calculated to balance the NH_4^+ budget.

The rate of NH_3 oxidation to NO_x, not discussed in Chapter 3, is taken from information published by Warneck (1988). This rate is insignificant. It is included in the figure only to illustrate that the cycles of reduced and oxidized nitrogen compounds are not entirely independent. The rates of other reactions are normalized to balance the cycle of the initial materials.

The total tropospheric mass of ammonium and ammonia is borrowed from the Warneck book. Warneck estimated these burdens by using the available results of surface and aircraft observations. It should be noted, however, that the total mass of ammonium ions seems to be too high compared to the tropospheric sulfate quantity presented in the next subsection. For calculating HNO_3 burden, the concentrations given in Table 4.1 are used. Moreover, we suppose a tropopause height at 10 km, and on the basis of aircraft measurements and model calculations, a constant HNO_3 concentration profile. This latter assumption is not entirely true since HNO_3 concentration increases slowly with increasing altitude in oceanic atmosphere, while over continents it decreases somewhat with height (Penner et al., 1991). The same is assumed for calculating the NO_x burden, which is also a bit questionable.

As observations show (Warneck, 1988) in the surface air, particulate nitrate concentrations on a mass basis are similar to the level of HNO_3 vapor. On the other hand, in the free troposphere HNO_3 occurs in higher concentrations than nitrate. For this reason one can assume as a first approximation that the magnitude of the nitrate mass is about half of that of nitric acid vapor. Obviously, the value received in this way should be considered with caution. Generally speaking, the accuracy of the burdens given in Figure 5.3 is not better than a factor of 2 or 3.

It follows from information collected in Figure 5.3 that the residence time of ammonia and ammonium in the troposphere is around 5 to 6 days. The corresponding figure for NO_x, HNO_3, and NO_3^- is equal to 2.4, 1.8, and 1.0 days, respectively. If these values are correct we can state that nitrogen species in a more oxidized state have shorter residence time in the troposphere. More research is needed, however, before the general acceptance of this conclusion.

Considering the fact that half of the ammonia molecules in the troposphere are man-made, it can be concluded that the atmospheric ammonium burden in the aerosol phase, as well as atmospheric NH_4^+, deposition, are controlled in a significant way by human activities. In the case of NO_x the anthropogenic fraction of the emission is even higher. Owing to short residence times human effects are concentrated in more populated continental regions where nitrogen species play an important part in the control of the acid depositions. This is illustrated in further detail by nitrogen budget calculations made for North America (Logan, 1983) and Europe (Bónis et al., 1980), showing an important transport from these continents to the air over other areas of the world.

5.3.2 The Tropospheric Sulfur Cycle

Since sulfur is the main component of fine aerosol particles and acid rain, the study of its cycle in the troposphere, including anthropogenic modifications, is of particular interest. Figure 5.4 summarizes our knowledge about the subject. It should be noted that Figure 5.4 does not contain the cycle of carbonyl sulfide important only in stratospheric chemistry. Further, the tropospheric pathways of sea salt particles and sulfur containing dusts (see Ryaboshapko, 1983) are not included since they are unimportant concerning the cycle of reactive gaseous compounds modified by human activities. Finally, we do not exclude the possibility that a part of reduced sulfur species, illustrated by DMS in the figure, is released in the form of H_2S and due partly to human activities. It is speculated, however, that DMS emitted mostly from oceanic sources dominates the tropospheric cycle of biogenic sulfur components. We also note in this respect that the value given for the global release of DMS in Figure 5.4 is the average of the emissions proposed by Andreae (1986) and Bates et al. (1987) (see Chapter 2).

Concerning information in Figure 5.4 it should be mentioned that the conversion terms for DMS are determined to balance the DMS cycle. The relative significance of the three possible reaction routes is taken as published by Warneck (1988). The formation rate of sulfate from sulfur dioxide is normalized by requiring the SO_4^{2-} reservoir to be balanced. One unit is then added to SO_2 dry deposition in Table 4.1 to maintain, in equilibrium, input and output terms of the sulfur dioxide cycle.

The tropospheric mass of MSA was calculated according to Warneck (1988) giving a column concentration of 25 μg S m^{-2} for oceanic air. The DMS burden was determined on the basis of the DMS emission and a residence time of 1.5 days (Andreae, 1986) by using Equation 5.2. Warneck (1988) compiled the results of aircraft measurements carried out to determine the vertical profile of SO_2 concentration. This compilation shows that in marine air the SO_2 concentration is independent of height, while in continental air it decreases in such a way that the surface concentration should be multiplied by 1250 m to obtain the column concentration. If we combine this information with SO_2 surface concentrations given in Table 4.1, a value of 0.60 Tg is obtained for SO_2–S burden, which is a bit higher than

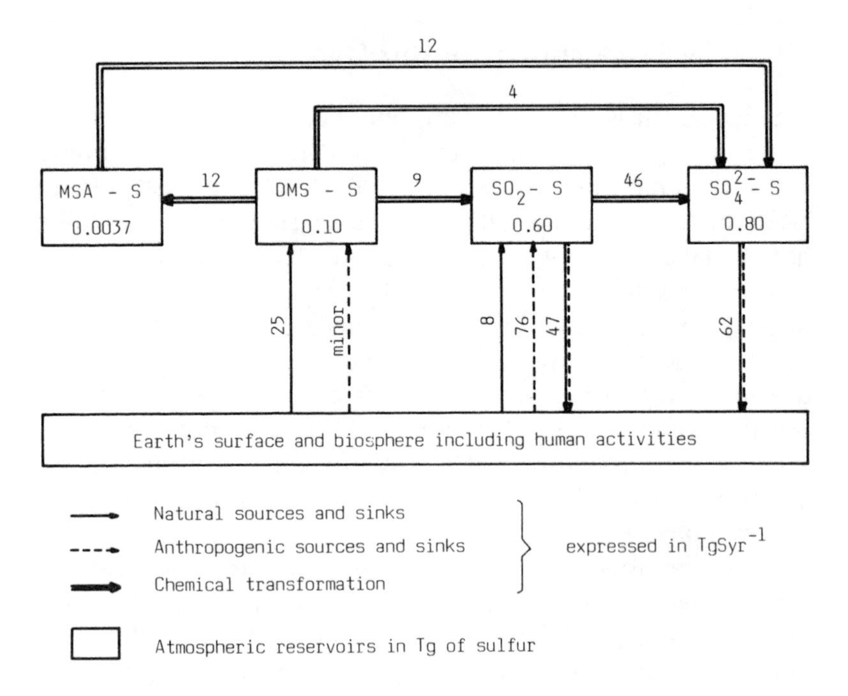

Figure 5.4 The troposphere budget of sulfur compounds. Note: MSA and DMS denote methanesulfonic acid and dimethyl sulfide.

the tropospheric sulfur dioxide mass calculated by Langner and Rodhe (1991). Finally, the tropospheric sulfate-sulfur quantity is taken from the work of Mészáros (1978), who calculated the SO_4^{2-} burden by using a similar approach to Warneck. We accept this relatively old figure since it agrees with the value received recently by Langner and Rodhe (1991) as a result of their model calculation.

The residence times of different sulfur species calculated on the basis of data in Figure 5.4 are given in Table 5.1. It follows from the data tabulated that the residence times, in contrast to those for nitrogen compounds, are longer for sulfur compounds in a more oxidized state. If we suppose that the transformation of SO_2 to SO_4^{2-} is a first order process, an overall conversion rate of $0.88\%/hr^{-1}$ can be calculated. This value agrees very well with the results of field studies aiming to determine this rate constant on the basis of atmospheric observations (Finlayson-Pitts and Pitts, 1986). The theoretical work of Langner and Rodhe (1991) suggest that the oxidation of SO_2 to sulfate takes place mostly in cloud water. These authors propose

Table 5.1 Residence Time of Different Sulfur Species in the Troposphere

Species	τ (days)
Methanesulfonic acid	0.11
Dimethyl sulfide	1.5
Sulfur dioxide	2.4
Sulfate particles	4.7

an annual SO_4^{2-} formation rate of 50 Tg S yr^{-1}. They calculate that 84% of this transformation is due to cloud oxidation of sulfur dioxide.

One can also see from Figure 5.4 that about 80% of atmospheric SO_2 comes from human activities. This means that 37 Tg SO_4–S formed annually from SO_2 is also man-made. Comparing this figure with the total sulfate formation rate of 62 Tg S yr^{-1} we can conclude that 60% of particulate sulfate in the troposphere originates from anthropogenic sources. This fraction is even higher if the budget of sulfur compounds is calculated for populated areas on a regional or continental scale (Galloway and Whelpdale, 1980; Mészáros and Várhelyi, 1982). Even the sulfur dioxide emitted over these continents contributes significantly to the sulfate burden over the North Atlantic Ocean and Asia. The importance of this conclusion, among other things, is clarified in the next chapter of this book.

6

Environmental Consequences of Anthropogenic Modifications: The Future of the Atmosphere

6.1 TROPOSPHERIC OXIDANTS

6.1.1 General Remarks

Generally speaking, we include into the class of tropospheric oxidants several chemical species, like ozone, peroxyacetyl nitrate, hydrogen peroxide, and oxygen containing free radicals (OH, HO_2); however, the oxidant concentration is often characterized by the O_3 level alone. As we have discussed in Chapter 3, chemical reactions in the troposphere produce a large amount of ozone. This O_3 formation was first identified in the Los Angeles area in photochemical smogs in the 1940s. Since that time the increase of the concentration of photochemical oxidants has become a regional and even worldwide phenomenon. Model calculations make it possible that 75% of tropospheric ozone is *in situ* product, while the rest comes from the stratosphere (Hough and Johnson, 1991). Two-thirds of tropospheric ozone are created from precursor gases emitted by man-made sources. More than half of anthropogenic ozone is due to industrial activity. The remaining part is the result of the reactions of gases released during biomass burning. Ozone

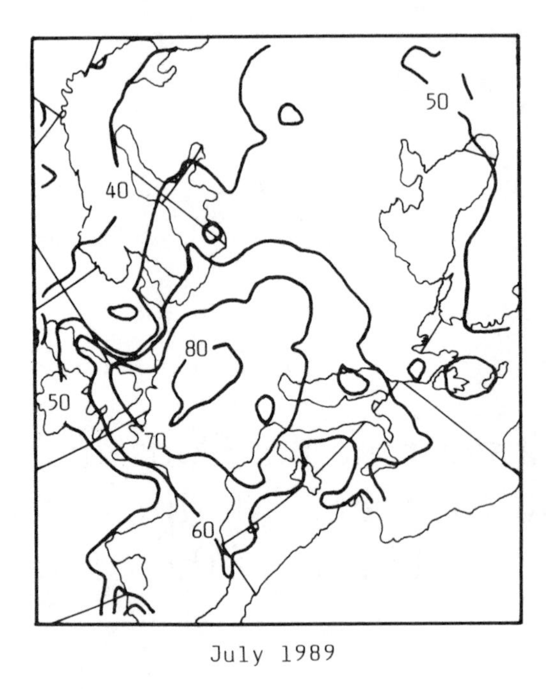

Figure 6.1 Calculated monthly means of daily maximum ozone concentrations expressed in ppb (Simpson, 1991).

produced by biomass burning is concentrated mostly in the Southern Hemisphere (Fishman et al., 1991).

As it was mentioned in Chapter 2, tropospheric ozone concentration has at least doubled in Europe during this century (Volz and Kley, 1988). The present increase is around 1% yr^{-1} as reviewed by Bojkov (see Penkett, 1991). The largest increase (1.6% yr^{-1}) has been detected at an elevation of 2000 m by ozone soundings made in Switzerland (Staehelin and Schmid, 1991).

6.1.2 Effects on the Biosphere

Ozone is a toxic gas for biological systems if its concentration in the air is above a certain level (~75 ppb) for a longer time. Such concentrations are very common in summer over Europe, North America, and Japan, mainly if the hourly maximums are considered. As an example Figure 6.1 shows the spatial distribution of the daily maximum O_3 concentrations over Europe as calculated for July 1989 by Simpson (1991). The

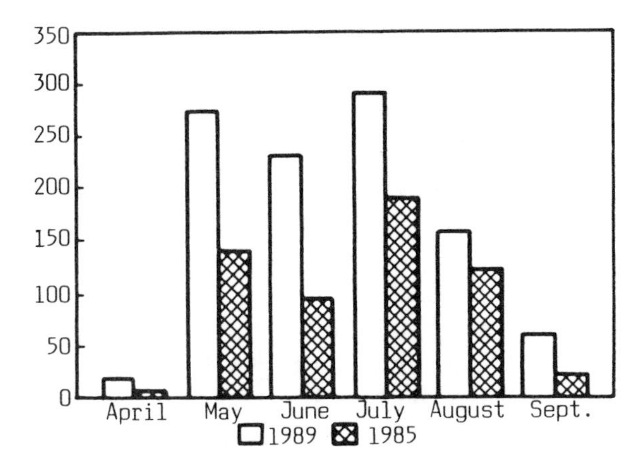

Figure 6.2 Calculated excess ozone for April–September 1985, 1989. Units: ppb hr per month (Simpson, 1991).

pattern shown in the figure agrees reasonably well with data measured at several points of the area. It can be seen that even the monthly average is above 75 ppb in certain regions of Western Europe; however, the individual hourly maximums can reach much higher values. If the ozone concentrations above 75 ppb ("excess" ozone) are multiplied by the hours of occurrence, the so-called exposure values are obtained. Figure 6.2 illustrates the results calculated in this way by Simpson (1991) for the summer part of the years between 1985 and 1989. These results clearly indicate the increase of the exposure in that time.

Exposure studies on humans show that in most healthy individuals serious irritation starts at 300 ppb (Heicklen, 1976). Fortunately, such high concentrations occur only on local scales under special conditions. Further, it is also demonstrated that persons exposed for a short time, even to 3000 ppb of O_3, recover rapidly when O_3 is removed. However, eye irritation already begins at oxidant levels of 100 ppb because of the presence of PAN and other photochemical oxidants.

Ozone also affects the vegetation by causing damages in crops and forests. Research indicates (Heck et al., 1982) that ozone is responsible for up to 90% of the crop losses in the U.S. caused by air pollution. The effects can be serious in particular if O_3 is combined with sulfur dioxide and nitrogen dioxide. Even in areas where the concentration has not exceeded the U.S. O_3 standard (120 ppb hr^{-1}) losses have been reported. The loss is estimated to be 2 to 4% of total crop production of the U.S. Experiments carried out with open top field chambers demonstrated that ambient O_3 concentrations observed in different parts of the U.S. produced a loss of 55 to 56%, 14 to 17%, and 10% in head lettuce, peanut, and soybean, respectively (losses were related to a control area with an O_3 concentration of 25 ppb). PAN also can create plant damages in concentration of 10 to 100 ppb. Thus, it was found that the injury of pinto bean and petunia is 55 and 30%, respectively, during 1 hr of exposure to a PAN concentration of 140 ppb.

As is well known, important forest decline has been observed in North America and Europe (Ashmore et al., 1985). The symptoms consist of premature defoliation and dieback leading to the death of the trees. In some regions, e.g., the southern part of Germany, the percentage of damaged trees exceeds 50%, including all tree species and all types of damage. It is speculated that ozone may be an important contributing factor to the phenomenon, since O_3 may destroy the leave cuticle, cell walls, and cell membranes of the leaves. In the Black Forest (Germany), where the relative damage was found to be above 50%, the annual and summer mean ozone concentrations measured from 1980 to 1983 were about 40 and 50 ppb, respectively. However, the annual maximum could exceed 150 ppb, and the percentage of summer hours with ozone concentration above 100 ppb varied between 1 and 5%. The authors referenced argue that the evidence that O_3 is the possible acting agent of damages is circumstantial; however, "ozone could conceivably be the prime cause of forest decline, it could be one important component of a combination of stresses, or it could be merely a secondary factor influencing trees already weakened by some other primary stress." Thus, it cannot be excluded that O_3 affects the trees in combination with other factors like the deposition of acidic cloud and/or precipitation elements as discussed later.

6.1.3 Changes in Oxidizing Capacity

The most serious consequence of the anthropogenic modification of tropospheric oxidant concentrations is the change in oxidizing capacity of the air. This leads to an important alteration of the rate of chemical reactions, which modifies the concentration and residence time of several compounds (Crutzen and Zimmermann, 1991).

Future changes can be calculated with appropriate numerical models (see Section 5.1) by using possible emission scenarios. Such an approach also makes it possible to estimate the effects of potential mitigation measures in the emission of different precursor gases.

A two-dimensional model was applied by Isaksen and Hov (1987) to determine the global effects of emissions changes in nitrogen oxides, carbon monoxide, and NMHC on the concentration of ozone and hydroxyl radical. The model calculations made for a time period between 1950 and 2010 indicate that if the methane emission were increased by 1.5% yr^{-1} the hydroxyl concentration would be reduced by 25% until 2010. During the same period the O_3 level would increase by 0.45% yr^{-1}. Further, a NO_x emission increase of 3% yr^{-1} alone would cause a significant increase in both O_3 and OH concentrations. Finally, if the emissions of CO, NO_x, and NMHC were increased by 3% yr^{-1} and the CH_4 release by 0.5% yr^{-1}, emission changes would result in an ozone increase of 1% yr^{-1} in agreement with the present observed tropospheric ozone trend mentioned above. More recently Crutzen and Zimmermann (1991) simulated tropospheric chemistry and its anthropogenic perturbations by means of a three-dimensional photochemical model. Their results show among other things that the tropospheric OH level has decreased considerably since the preindustrial time. The concentration decrease during daytime reaches 20% at mid-tropospheric altitudes over the Equator and in the entire troposphere over about 30° N. These results indicate the importance of tropical regions in tropospheric chemistry. On the basis of their modeling the authors propose that the global average concentration of hydroxyl radicals in the troposphere is presently equal to 7×10^5 molecules cm^{-3}.

The future regional effects of increasing anthropogenic emissions of precursor gases on tropospheric oxidant concentrations was recently determined by Thompson et al. (1991). These authors carried out their calculations by means of a one-dimensional model. More exactly, they calculated the vertical profile of the concentration of different oxidants over "coherent" regions determined on the basis of combustion NO emission. In this way Thompson and her associates divided northern

Table 6.1 Regional Ozone Concentration Increase Foreseen until 2035 with Increasing Methane Concentration (0.8% yr^{-1}) and Carbon Monoxide Emission (0.5% yr^{-1})[a]

Region	1985 (ppb)	2035 (ppb)	Increase (%)
Urban mid-latitude	59	70	19
Continental mid-latitude	42	48	14
Marine mid-latitude	24	26	8
Marine low-latitude	17	19	12
Continental low-latitude	22	25	14
Southern mid-latitude	25	27	8
Area-weighted total	24.4	27.4	13

[a] According to Thompson et al., 1991.

mid-latitudes into three regions (urban, continental, and marine) and low-latitudes in the Northern Hemisphere into two. A region representing the oceanic areas at mid-latitudes of the Southern Hemisphere was also considered. For the calculations concentration values relative to the year 1985 were used as input data. The model was run until 2035 by assuming different emission scenarios. In one of the scenarios it was assumed that the methane concentration rises by 0.8% yr^{-1} and the carbon monoxide flux increases by 0.5% yr^{-1} over all regions. The NO emission was held constant. The results obtained for O_3 are tabulated in Table 6.1. One can see that the ozone concentration will rise over all regions studied. The change in ozone level is predicted to be higher over continental areas, mostly under more polluted conditions (urban region). The area-weighted average increase is equal to 13% for the whole period. Model runs also indicate that between 1985 and 2035 the level of H_2O_2 and HO_2 will increase by 22% and 8%, respectively, for the same scenario. Owing to the rise of the concentration of the latter two species the level of OH in the troposphere will decrease globally by 13%. Thus, there will be a loss of gaseous oxidizing capacity in the troposphere with an increase in aqueous-phase oxidizing potential. Among other things this will intensify the wet removal of sulfur dioxide leading to the fall of more acidic precipitation.

The effects of the possible reduction in European volatile organic compounds (voc) and nitrogen oxide emissions on the

Table 6.2 Relative Decrease (%) in Averaged Concentration of O_3, NO_x, and PAN at One Receptor Point in the Netherlands (Delf) in the Case of Different European Emission Reductions (%) of NO_x and VOC Emissions[a]

	Emission Reduction				
Run	NO_x	VOC	O_3	NO_x	PAN
1	30	40	3.5	28	35
2	50	40	2.7	48	37
3	50	70	6.3	47	59

[a] Taken from data of De Leeuw and Van Rheineck Leyssius, 1991.

oxidant concentrations in the Netherlands were studied by De Leeuw and Van Rheineck Leyssius (1991) by Lagrangian type model calculations.* These authors assumed a constant background CH_4 concentration and did not change the natural release of NO_x and VOC and the total CO emission. They estimated the initial concentrations on the basis of the work of Isaksen and Hov (1987) discussed above. Their results obtained in three different scenario runs are given in Table 6.2. Data tabulated indicate that the decrease of the O_3 level is relatively unimportant even in the case of very significant reductions in NO_x and VOC emissions. At the same time NO_x and PAN levels are lowered significantly owing to the fact that the O_3 production per one NO_x molecule is lower at low NO_x concentrations (Lin et al., 1988). Thus, we can conclude that the control of tropospheric ozone concentration is not an easy task. The best way would be to reduce significantly the VOC emissions. However, VOCs are partly of biological origin even over the populated continents. Thus, the question needs further considerations.

Finally, it should be noted that ozone molecules absorb infrared radiation emitted by the Earth's surface. This means that increasing tropospheric ozone concentration contributes to the anthropogenic greenhouse effects discussed in Section 6.4.

* In this approach air columns of unit base move along the wind trajectory above the emission field. The height of the columns is equal to the mixing height. For each time step the concentration variations in the columns are calculated by considering the emission, removal, and chemical transformations.

6.2 CONSEQUENCES OF CHANGES IN ATMOSPHERIC DEPOSITION

6.2.1 The Problem of Acid Deposition

The definition of "acid rain" is not self-evident. The complication is caused by the fact that even in clean air, free from human influences, there are acid aerosol particles and gases that lower the pH of cloud and precipitation waters below the atmospheric neutral point defined in Section 4.2.2. Thus, it can be demonstrated (Charlson and Rodhe, 1982) that atmospheric sulfur compounds of biogenic origin can decrease the pH until 4.5. This is true in particular under remote oceanic conditions where the pH of atmospheric water is controlled by acidic sulfur containing particles formed from DMS (Vong et al., 1988). On the other hand, as recent studies show (Andreae et al., 1988), in tropical areas organic acids formed from biogenic precursors control significantly the acidity of the atmosphere. It follows from these arguments that the distinction between natural (biogenic) and anthropogenic acidities is not always easy. Figure 6.3 gives an idea of the average pH values of precipitation observed under different conditions (Miller, 1984). It can be seen that in remote areas the pH is around 5.0. Hence, one can propose with caution that this value represents in first approximation the natural average pH, which is lowered in polluted air (northeastern U.S.) by nearly one order of magnitude by sulfuric and nitric acids. The contribution of sulfate ions in acidity control is three times more important than that of nitrate ions as the lower part of Figure 6.3 suggests.

For illustrating in more detail the distribution of pH in precipitation over continental regions, the results gained in North America are plotted in Figure 6.4 (Barrie and Hales, 1984). Generally, low pH values occur in such regions (northeastern U.S., southeastern Canada) where sulfur dioxide and nitrogen oxide emissions are significant. In these areas, similar to the situation in many parts of Europe (Nodop, 1986), precipitation water is very acidic. Over other regions of North America the emission of SO_2 and NO_x is smaller, and ammonia gas and dust particles also contribute to the control of hydrogen ion concentration. Hence, beside energy production, agricultural and soil sources also play a role in pH control.

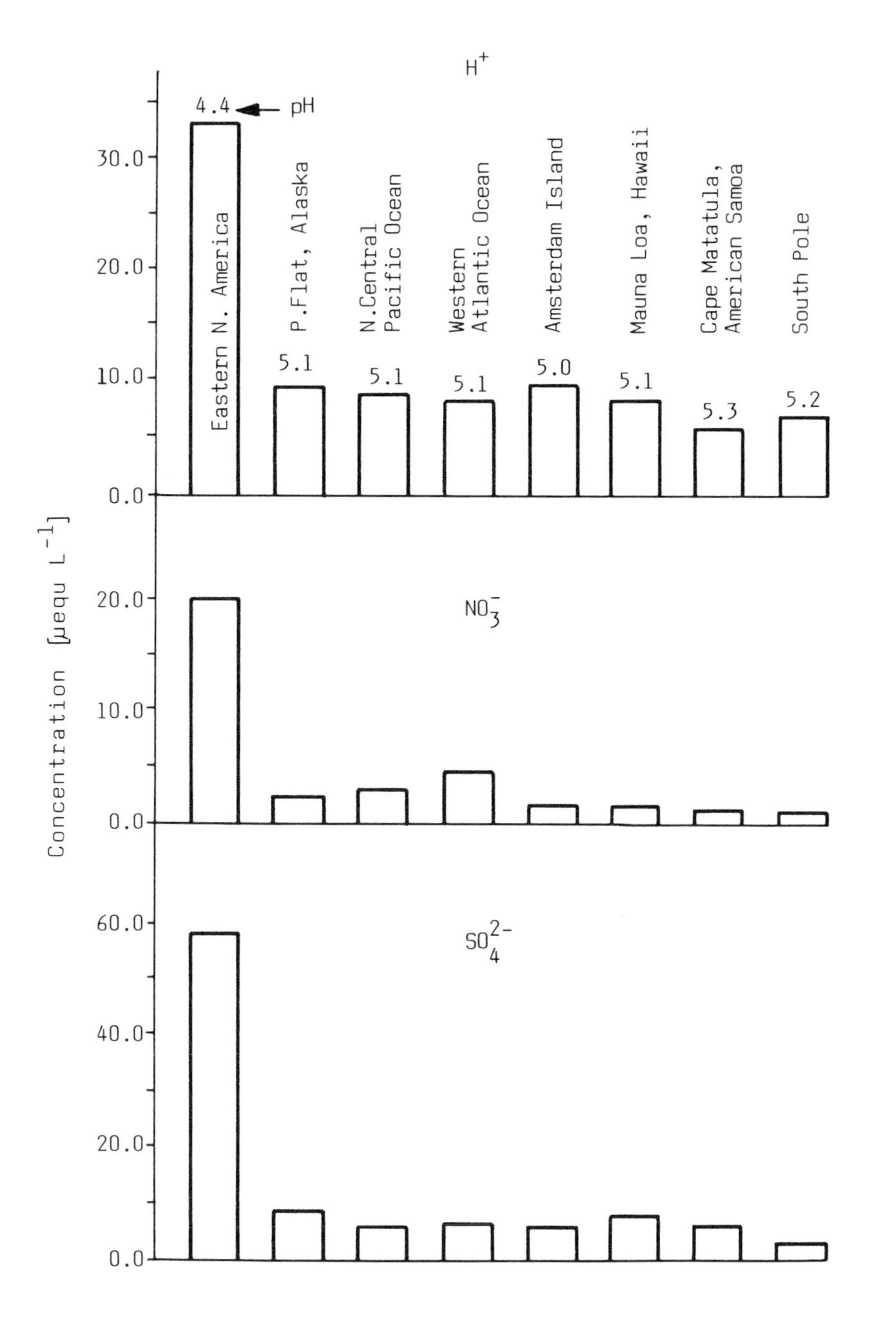

Figure 6.3 Acidity and composition of precipitation under different conditions (Miller, 1984).

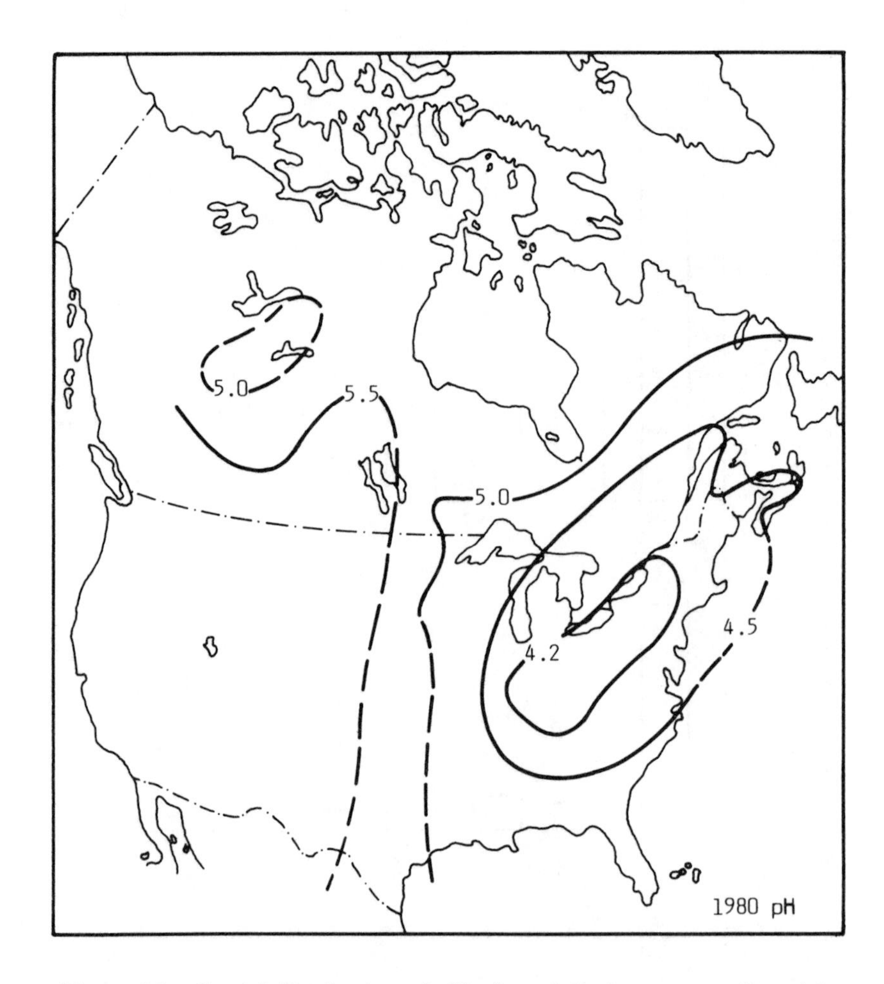

Figure 6.4 Spatial distribution of pH of precipitation water collected in North America (Barrie and Hales, 1984).

Turbulent downward transfer of gases and aerosol particles further increases the acidity of atmospheric deposition. In this respect the role of sulfur dioxide, nitrogen dioxide, and nitric acid vapor should be stressed in particular. On the other hand, at higher elevations in mountainous areas the interception of acidic cloud droplets by tree leaves dominates the magnitudes of acid depositions (see Mészáros, 1992).

Acid depositions affect both terrestrial and aquatic ecosystems. The impact on aquatic ecosystems is serious, particularly in lakes with low buffering capacity. Such lakes are located in areas where rocks and soils are composed of gneiss and granite or where they are poor in lime. Such lakes can be found, e.g., in Scandinavia (where the problem was first recognized), the Adirondack Mountains (U.S.), and Nova Scotia (Canada). The ecosystem of these lakes is very sensitive to any changes in hydrogen ion concentration. If the pH is lower than 5.0, acidity can directly cause mortality of some fishes. Further, if the pH decreases until about 4.5 insoluble aluminum compounds become soluble. The aluminum ions formed have a toxic effect on many aquatic organisms, and fishes die because of aluminum poisoning. Under these conditions the water of the lakes is very clear and only white moss lives in the water in larger quantities. It should be noted that not only the hydrogen ion concentration of surface waters can be altered. There is a large body of evidence suggesting that atmospheric depositions have also raised the acidity of groundwaters, which have become corrosive, and have leached metals from soils, water tanks, and pipes to drinking water (for further details see Swedish Ministry of Agriculture, 1982).

Another important regional consequence of acid deposition is the forest decline. The hypotheses for explaining the impact of acid deposition on forest ecosystems can be divided essentially into two categories:

1. Direct foliar effects
2. Indirect effects through soil acidification

In the first case acid deposition damages directly the protective layer outside the leaves, leaches out nutrients (e.g., magnesium), and finally increases the vulnerability of trees to pests and disease. This effect can be important in particular if leaves have been already affected by gaseous pollutants like ozone. The second hypothesis is based on the observation that acidic deposition increases the hydrogen ion concentration in forest soils. Higher acidity, as in the case of lakes, leads to the mobilization of heavy metals including aluminum. Root hairs of the trees are destroyed by soluble aluminum, which makes the water and nutrient uptake impossible. Although similar effects can be produced in agricultural soils, they are less probable since the pH of agricultural lands is controlled essentially by the quantity and quality of fertilizers applied. However, even in this case rains with low pH may cause temporary stresses.

This short discussion clearly indicates that the control of the emission of acidic substances is of crucial importance. Besides

SO_2 and NO_x this control should be extended to the release of ammonia, since this species, after being deposited onto the soils as ammonium, produces hydrogen ions by nitrification processes (Buijsman and Erisman, 1988). The need for such mitigation procedures is rather recent since damages generally occur rather far from the sources and their origin has not been evident in the past. Moreover, for acid deposition control international efforts are necessary due to transboundary air pollution transport. For this reason international cooperations, e.g., the European Monitoring and Evaluation Programme (EMEP) of the Economic Commission for Europe, are needed to elaborate the basis of continentwide air quality management.

Another difficulty is caused by the fact that the relationship between source strength and deposition distribution over a larger area is rather complicated owing to the complexity of the meteorological, chemical, and removal processes involved. This means that suitable numerical models are needed to relate sources to deposition and environmental impacts. Such a model, called RAINS (Regional Acidification Information and Simulation), was constructed by the acid rain group of the International Institute for Applied Systems Analysis (Laxenburg, Austria) to study present and future acidification patterns in Europe as a function of the energy policy. The atmospheric part of RAINS (Alcamo et al., 1990) is a Lagrangian model taking into account both dry and wet deposition. The circulation of sulfur compounds is simulated by the EMEP model developed by Eliassen (1978), while for the calculation of NO_x–N deposition the model originally proposed by Derwent (1986) is applied. Finally, NH_3/NH_4 deposition is calculated by the model of Asman and Janssen (1987). By running the model the European acid deposition distribution was determined (see Alcamo et al., 1990) for 1985 and 2000 in the case of three emission scenarios elaborated by Amann (1990). The first scenario ("No Controls": NC) is theoretically the worst case: the total European energy consumption increases from 103 EJ (1985) to 140 EJ (exajoules = 10^{18} J) until 2000, but the SO_2 emission (~27 Tg S yr^{-1}) is practically stabilized due to the continuing conversion to nuclear power. In contrast to this, NO_x emissions are expected to increase until 2000 (from 10.5 to 14 Tg N yr^{-1}) owing to the transformation sector.

In the second scenario run ("Current Reduction Plans": CRP) it is assumed that the countries change their SO_2 and NO_x emissions according to their commitments in international conventions* and through national abatement plans (unfortunately not each country announced such a commitment). This reduces the SO_2 emission by 18% and the NO_x release by 5%. Finally, in the third scenario ("Best Available Technology": BAT) for large boilers (power plant, industry etc.) a flue gas desulfurization with more than 90% efficiency is applied, and it is supposed that in small boilers (mainly in the domestic sector) only low-sulfur fuel oil (with 0.15% S content) is used. According to this scenario all existing technically feasible controls come into operation to reduce the NO_x emissions. These control strategies lead to SO_2 and NO_x emission reductions of 82 and 55%, respectively. For the orientation of the reader we note that the annual cost necessary for the SO_2 reduction in the CRP scenario amounts to a sum of 12 billion DM (~8 billion U.S. dollars), while the corresponding figure for BAT is estimated to be 85 billion DM (~57 billion U.S. dollars).

Figure 6.5 illustrates the European sulfur depositions calculated for 1980 and 2000. In 1980 deposition levels in Central Europe exceed the value of 8 g S m^{-2} yr^{-1}. For over 62% of the area of the continent the deposition is higher than 1 g S m^{-2} yr^{-1}. The deposition distribution calculated by NC scenario for 2000 is quite similar to that obtained for 1980, while for the CRP scenario the area with high depositions is slightly decreased. In the latter case the area having depositions greater than 1 g S m^{-2} yr^{-1} is decreased to 53%. In the case of the BAT scenario important changes can be noted if we compare the pattern in 1985 and 2000. The maximum depositions in 2000 are between 2 and 4 g S m^{-2} yr^{-1}, and in a large part of Europe the sulfur deposition is reduced to 0.5 g m^{-2} yr^{-1}. Finally, in 2000 the area with depositions exceeding 1.0 g S m^{-2} yr^{-1} is only 5% in contrast to the values of 62% in 1980.

Similar calculations were made by the IIASA group for characterizing nitrogen deposition. In these model runs the NO_x emissions are supposed to be controlled until 2000 as

* According to the "Sulfur Protocol" the countries which signed the document will reduce their SO_2 emission by 30% until 1993.

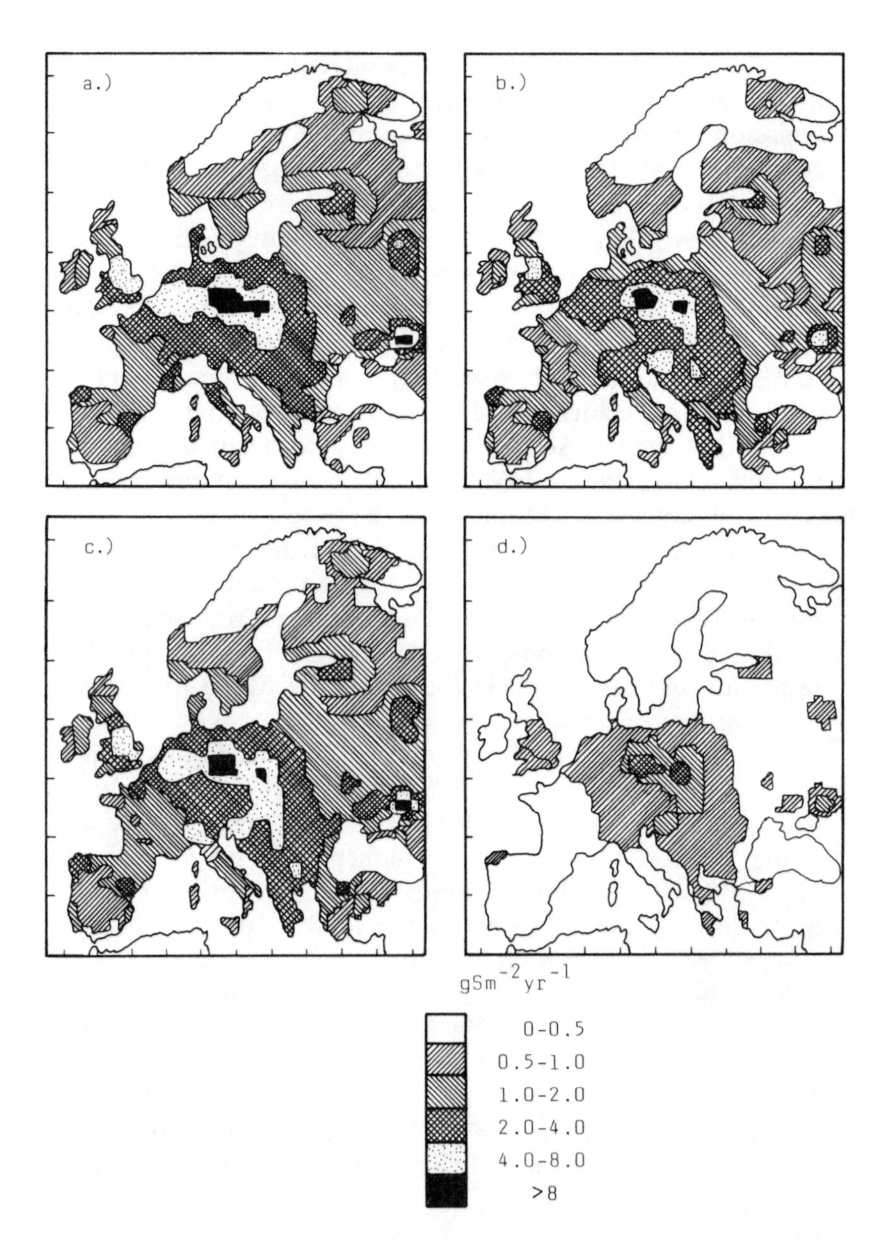

Figure 6.5 Calculated deposition of sulfur in Europe (Alcamo et al., 1990) a: in 1980; b: in 2000 using "Current Reduction Plans" scenario (see the text); c: in 2000 using "No Control" scenario; d: in 2000 using "Best Available Technology" scenario.

discussed previously for different scenarios, while the NH_3 emissions were kept at the 1980 levels. The results of these simulations show that the area having N depositions higher than 1 g m^{-2} yr^{-1} is reduced from 52% (1980) to only 41% (2000) even in the case of the BAT scenario. According to the results of calculations it is expected that in acid depositions the importance of nitrogen relative to sulfur will increase in the future. RAINS model contains a submodel for soil acidification based on the conversion of sulfur deposition into H^+ load in soils (Kauppi et al., 1990). One can estimate by running this submodel that in the relative area of Central European forest soils with pH <4, NC and CRP scenarios in 2000 will be 25 and 20%, respectively. By applying the emission mitigation in Europe according to the BAT scenario no such areas will be found.

On the basis of the discussion in this subsection one can conclude that acid deposition has serious impacts on our environment on a regional scale. Model calculations show that this situation may be improved by applying appropriate emission abatement policy; however, for the realization of such a policy important capital investment and satisfactory international conventions are needed.

6.2.2 Heavy Metals

As we have seen previously several toxic metals are emitted by human activities into the atmosphere. These metals are generally associated with fine aerosol particles and are transported far from emission sources to produce regional and even global problems. Consequently, toxic metals cause not only health hazards by inhalation in densely populated areas, but the increase of their concentration in the air also leads to ecological impacts in the remote environment. Atmospheric trace metals, being deposited to the biosphere in large areas, modify the natural uptake of microelements by different plants. In this way metals accumulate in plants and plant-eating animals, and they disturb finally the food chain of humans.

In the human and animal body, toxic metals can cause several problems. Elevated levels of cadmium cause kidney disease, gradual accumulation of lead can poison the entire body, including the nervous system, the digestive tract and blood-forming tissues, while vanadium inhibits the synthesis of cholesterol and other lipids. Zinc can produce either lung or intestinal tract manifestation, copper in excess accumulates in the liver and is associated with the induction of hemolytic disease, and finally, nickel can cause lung cancer of certain animal species (Dvorak and Lewis, quoted in Pacyna, 1981).

It follows from this discussion that the study of atmospheric occurrence and deposition of trace metals is of crucial importance (Barrie and Schemenauer, 1989; Puxbaum, 1991). On the other hand, the modeling of the long-range transport of trace metals in the atmosphere is also of interest, since transport models link emission strength and distribution with atmospheric concentration and deposition. Such model studies have been made recently by several workers for different parts of Europe (Pacyna et al., 1989; Borbély-Kiss et al., 1991) and for the North and Baltic Seas (Petersen et al., 1989).

Due to these efforts the present situation is rather well known, mostly for Europe. Obviously, the future environment depends on emission regulations applied by different countries. Pacyna et al. (1991) made emission calculations for the year 2000 by assuming that in Western Europe all electric power plants will comply with the German regulation accepted for new power plants. On the other hand, in East European power plants German regulations for existing plants will be applied. They also assumed that the industrial release of different heavy metals will be regulated according to the best available technology. Further, gasoline in Western Europe will be totally unleaded, while the lead content of gasoline in Eastern Europe will be 0.15 g L^{-1}. On the basis of this scenario the authors cited above calculated that in 2000 the European lead emission will be a third of the value corresponding to 1982. Calculations also showed that the relative reduction of cadmium release in the 1990s will be around 0.5. This means that important emission decreases are foreseen; however, for these reductions the mitigation measures mentioned should come in force.

6.3 THE STRATOSPHERIC OZONE HOLE

6.3.1 The Discovery of the Antarctic Ozone Hole

The discovery of the spring ozone hole over Antarctica (Farman et al., 1985) is a milestone in the history of atmospheric ozone research. On the one hand, the discovery of the ozone hole first demonstrated the consequence of the use of chlorofluorocarbons (CFCs) and led to the recognition that ozone chemistry cannot be understood if heterogeneous reactions are ignored (see Section 3.2.2). The observational evidences summarized below indicated that the ozone cycle in the stratosphere is even more complex than we believed in the 1970s after many modifications of the theory.

For the illustration of this statement let us consider Figure 6.6 (Schoeberl et al., 1989) showing the results of total ozone observations carried out from the ground (Halley Bay, Antarctica) and from satellite platform in a spring month (October) over Antarctica. It follows from the curve based on ground measurements that in the 1950s and 60s the value of total ozone varied around 300 DU (Dobson unit, 1DU = 10^{-3}cm). Since that time the ozone quantity in an air column decreases owing to the increase of the concentration of chlorine containing species formed from anthropogenic CFCs (see, e.g., Prather and Watson, 1990). As we have seen in Chapter 3 the reason for the phenomenon is the release of active chlorine by ice crystals in which heterogeneous reactions occur. The efficiency of these reactions depends on the meteorological conditions as illustrated by the two-year cycle in data caused by the quasi biennial oscillation of stratospheric circulation. This means that the dynamic characteristics (and consequently the temperature) of the stratospheric vortex, observed regularly over Antarctica in wintertime, influence the ozone hole formation as well.

One can see from Figure 6.6 that the spring Antarctica ozone hole does not mean a real "hole". It means that the total ozone can decrease ~50% below its normal value, which cannot be explained by homogeneous chemistry alone (see later Figure 6.8a), even if the effects of CFCs are taken into account. Measurements made by ozonesondes indicate that this ozone decrease is due to the depletion of O_3 molecules between 15 and 20 km (Watson, 1989) where it can be as great as 95%. Satellite observations also provide a good possibility to determine the horizontal extension of the ozone hole. Figure 6.7 gives the hemispheric distribution of total ozone obtained by satellites during October of the years 1985 through 1988 (Schoeberl et al., 1989). On the basis of this information one can conclude that the dimension of the ozone hole can be very significant and can reach populated areas like Australia as

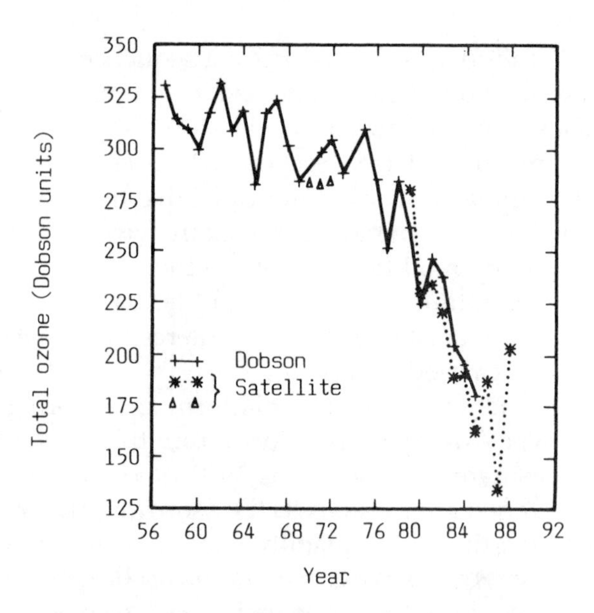

Figure 6.6 The monthly mean October total ozone values over Halley Bay as measured by Halley Bay Dobson and satellite instruments (Schoeberl et al., 1989).

discussed in more detail by Atkinson et al. (1989). It should be noted, however, that its occurrence is limited to about two months of the year.

6.3.2 The Case of the Northern Hemisphere

Above the North Pole the winter stratospheric vortex is less pronounced than over the South Pole because of dynamic perturbations caused by the formation of planetary waves over high mountains (e.g., Rockies, Himalayas) in the Northern Hemisphere. Consequently, mixing with the air over lower latitudes is more intensive, and except unusual situations, the temperature at proper stratospheric heights is somewhat higher than over Antarctica. At higher temperatures the formation of polar stratospheric clouds consisting of ice crystals is obviously less probable.

The first measurements of chlorine species in the Arctic winter stratosphere were carried out in 1988 above Thule, Greenland (Solomon et al., 1988). These studies showed that nighttime chlorine dioxide concentrations were about 10 times higher than theoretical predictions based on homogeneous photochemical theory. At the same time (ClO_2) abundances

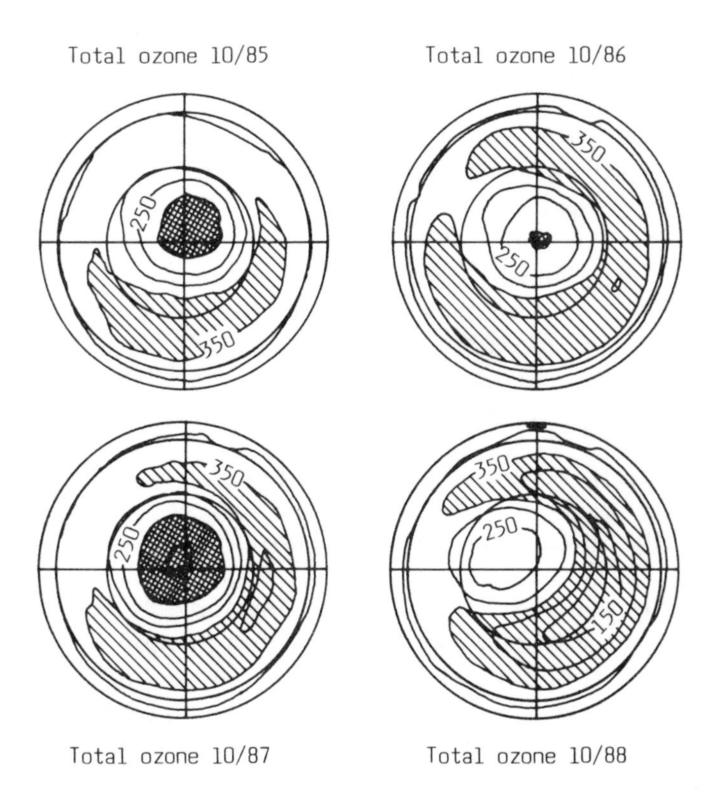

Figure 6.7 Orthographic, South Pole centered, October monthly mean total ozone for the years 1985, 1986, 1987, and 1988 on the basis of satellite data. The Greenwich meridian appears at 12 o'clock in each figure; the circles indicate 30 and 60°S latitudes. Lightly shaded areas show values greater than 350 DU (Dobson units); dark shading, values less than 200 DU. Contour increment is 50 DU (Schoeberl et al., 1989).

were lower than the corresponding values for Antarctica by a factor of 5. Observations also revealed an O_3 depletion that was relatively modest (5 to 10%) compared to the Southern Hemisphere. Hence, it was concluded that over the Arctic there is a chemical "preconditioning" for the ozone hole, but meteorological conditions hinder its development. However, further observations made in 1989 over the Arctic revealed active chlorine concentrations comparable to the levels characterizing the Antarctic situation (Brune et al., 1990). In this respect we refer to Figure 3.2, showing that in the Northern Hemisphere the highest negative trend (~1% yr^{-1}) in total ozone occurs over mid-latitudes in winter months. Therefore, one cannot exclude the possibility that this ozone depletion is due to the transport of chemical precursors to mid-latitudes where solar radiation is available (Brasseur, 1991). This means that chlorine species are activated over a much larger area than in the Southern Hemisphere. However, the role of stratospheric sulfate particles in the process (see Section 3.2.2) should be clarified by further research.

Briefly, a potential exists in the stratosphere over high latitudes at the Northern Hemisphere for the formation of an ozone hole. Since this potential will increase in the future even in the case of the total phaseout of CFCs (see the next subsection), the continual monitoring of ozone and related species is an obvious task for the scientific community.

6.3.3 Model Predictions for the Future

Another important duty of scientists is to forecast future changes in the ozone layer on the basis of the amount of CFC emissions planned. This can be done by applying appropriate chemical models containing all important reaction steps of the ozone formation and destruction. Before presenting the results of such model simulations we note that the present $CFCl_3$ and CF_2Cl_2 release is 0.36 and 0.45 Tg yr^{-1}, respectively. These two species are responsible for more than half of the chlorine load in the stratosphere, while 20% of chlorine is formed from methyl chloride of natural origin (Prather and Watson, 1990). It is estimated that 70 to 90% of the ozone depletion over spring Antarctica is due to chlorine, the remaining part is destroyed by bromine coming into being by the photolysis of Br containing halocarbons (halons) emitted by man.

Figure 6.8 shows predicted total ozone changes for the time interval between 1960 and 2010 as a function of latitude and time. For illustrating the differences between the roles of homogeneous and heterogeneous chemistry, these cases are separately presented. Thus, in the calculations given in Figure 6.8a homogeneous chemistry was taken into account and the authors (Brasseur et al., 1990) assumed, as in the case of the other parts of the figure, a 95% phaseout of CFCs by the year 2000. One can see that ozone depletions are modest, they reach maximum at 1% at 85° N and 1.5% at 85° S in late winter and early spring (compare this latter with the ozone depletion observed during the Antarctic ozone hole, Figure 6.6). Even positive changes are predicted owing to the cooling

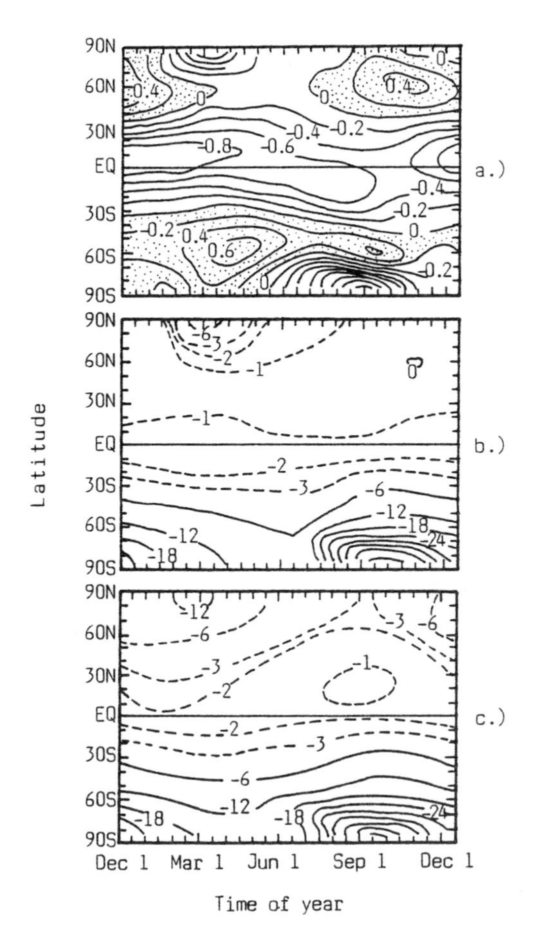

Figure 6.8 Ozone column changes (%) as a function of latitude and season calculated between 1960 and 2010 when (a) only homogeneous chemistry processes are taken into account, (b) heterogeneous processes occurring on the PSCs in the polar regions are introduced, (c) heterogeneous processes on both PSCs and sulfate aerosols are considered (Brasseur et al., 1990).

of the stratosphere* for compensating global warming and to an enhanced ozone production in the troposphere (see Section 6.1.2). The depletion values are completely different if heterogeneous processes occurring in polar stratospheric clouds are included into the model. The pattern in Figure 6.8b indicates clearly the spring Antarctic ozone hole, and the

* The rate constant of thermal chemical reactions depends on the temperature.

calculations for 2010 predict a spring Arctic ozone reduction of about 6%. Moreover, when the effects of stratospheric aerosol particles are also considered (Figure 6.8c), the model predicts further ozone depletion mainly during winter and spring months over the Northern Hemisphere. Finally, all the calculations indicate that the minimum O_3 reduction (~1%) is foreseen for the equatorial stratosphere.

Our discussion above shows that chlorine plays an important part in anthropogenic ozone depletion in the stratosphere. It is estimated (Prather and Watson, 1990) that stratospheric chlorine concentrations below 2 ppb are needed for the recovery of the Antarctic ozone hole. Among other things this estimation is based on the fact that in the late 1970s, when the ozone hole appeared, 2 ppb was a general chlorine level in the stratosphere (the present value is 3 ppb). Prather and Watson (1990) calculate that even if the emission of all halocarbons were eliminated by the end of the years 1995, 2000, and 2005, the chlorine level of the stratosphere would decrease to a concentration of 2 ppb only in 2055, 2073, and 2091, respectively. In other words, each five-year delay in the phaseout of these chemicals lengthens the occurrence of the ozone hole by about 18 years. This means that urgent international steps would be necessary to mitigate stratospheric environmental impacts. The Montreal Protocol (1987) aiming to regulate the use of CFCs at an international level is certainly not sufficient since it predicts a decrease of only 50% until July 1998.

Another possible way* examined for the reduction of the impacts is to replace present CFCs by other halocarbons. The philosophy of the substitution is to utilize species having shorter residence times and less chlorine atoms. While the use of such substitute compounds could somewhat reduce human influences, one can declare with Prather and Watson (1990) that "Ideally, we should cease emissions of CFCs and other halocarbons immediately. All other options result in enhanced levels of stratospheric chlorine and bromine sometime in the future." This conclusion states that the solution of the problem is in the hands of chemical manufacturers and politicians. It is

* In a recent paper Cicerone et al. (1991) proposed to reduce the ozone hole depletions by injecting alkanes into the Arctic stratosphere and removing chlorine atoms by chemical reactions. However, further research is needed to investigate this possibility.

hoped that they accept scientific arguments and will act in the interest of future generations of mankind.

6.3.4 Effects of Increased UV Radiation

As we have seen in Chapter 3 ozone in the atmosphere provides a shield against UV radiation coming from the Sun. The UV part of the solar spectrum consists of three wavelength bands: wavelengths from 320 to 400 nm are termed UV-A, wavelengths from 280 to 320 nm are called UV-B, and wavelengths from 200 to 280 nm are known as UV-C. Radiations in the UV-C band are totally absorbed by the atmosphere, while UV-A is reflected partly by O_2 and N_2 molecules. The interesting range is UV-B, since its absorption depends on the ozone content of the atmosphere. It is speculated that a 10% ozone decrease would result in an increase of 20% of the intensity of UV-B radiation at the surface (WMO, 1976). Since UV-B radiation affects the biosphere, the study of its changes at the surface induced by ozone depletion is of particular interest for environmental considerations (see, e.g., Titus and Seidel, 1988 and other papers in the same volume).

The most important environmental issue is that UV-B is responsible for sunburn and can cause skin cancer, mostly melanoma. However, the relationship between UV intensity and melanoma is not absolutely clear; it depends on several factors like the exposure to solar radiation, individual susceptibility, and personal behavior. Epidemiological studies indicate that the increase of UV radiation can also lead to cataracts, retinal disorders, and herpes infections.

Laboratory experiments carried out in growth chambers reveal that many plants are sensitive to UV-B radiation. It is found that peas, beans, squash, melons, and cabbage seem to be the most sensitive. The potential impact can be especially important if UV-B exposure is combined with water stress or mineral deficiency. Experiments show that a 25% depletion in ozone results in a 20 to 50% reduction in soybean yield, at least under laboratory conditions. Field studies showed less sensitivity of plants to UV radiation. This means that the problem is far from solved in a satisfactory manner.

Finally, aquatic biosphere can be adversely affected by intense UV-B radiation. This is true in particular for phytoplankton spending their time in surface water. If their productivity were reduced by ozone depletion, this would produce serious effects, because these plants serve as food for almost all fishes. Even the larvae of many fishes, important as protein sources for people, would also be damaged.

Without discussing this question in more detail, we note that for the estimation of the impacts of stratospheric ozone depletion the effects of the possible increase of tropospheric ozone must be taken into account simultaneously. As we have seen, the increase of tropospheric ozone can be rather important over populated regions in mid-latitudes, especially in the summer half-year. Thus, Brühl and Crutzen (1989) calculate by an appropriate model that, owing to human activities, the UV radiation intensity could even decrease at the surface in summertime, mostly if the effects of aerosol particles and clouds are also considered. On this basis Penkett (1989) speculates that the main consequences of man on ozone is not the intensification of UV radiation at the surface but the change in atmospheric structure and dynamics due to the stratospheric decrease (less UV radiation absorption) and tropospheric increase (more infrared absorption) of the temperature. The differences among opinions obviously suggest that further studies should be done to determine the environmental impacts of anthropogenic modifications of atmospheric ozone.

6.4 CLIMATIC EFFECTS OF ATMOSPHERIC GREENHOUSE GASES

6.4.1 Bases of Climate Modeling

The Earth's climate is the result of the interaction of several terrestrial and extraterrestrial factors (see Chapter 1). Terrestrial factors include the composition and thickness of the atmosphere, the area and distribution of oceans, continents, and ice cover, as well as the nature and extension of the biosphere. On the other hand, external factors comprise parameters like the Sun-Earth distance, the Sun luminosity, and different characteristics of the Earth's orbit. Variation in any factor mentioned raises the possibility of climate change. Thus, if human activities modify the composition of the atmosphere climate changes are foreseen. The evaluation of such changes is rather complicated because climate is a nonlinear system. Further, a given modification of the composition induces so-called feedback mechanisms, which are not always self-evident. Briefly, the numerical description of the effects of the variation of initial parameters is not easy, and it is possible only by complex climate models.

Generally speaking, climate models are composed of mathematical representation of physical laws controlling the climate system behavior. The physical laws include Newton's second law, the principle of conservation of mass, the first law of thermodynamics, the laws of radiative transfer, and the principles of diffusion and phase transition of water vapor. The mathematical formulation of these principles gives a closed system of equations governing the model. In the governing equations we can find such parameters as the horizontal and vertical wind components, the atmospheric pressure, temperature and density, as well as the mass fraction of different chemical substances. For the solution of model equations the initial boundary conditions (e.g., external forces) should be specified and the calculation is made by integration for the time period needed. Since atmospheric processes take place on very different temporal and spatial scales, an important part of climate modeling is the consideration of the interaction of macroprocesses with phenomena taking place on the micro scale (<100 km) between the grid points of the calculation (radiation transfer, turbulence, process of cloud physics, and air chemistry). Phenomena of smaller scale can be included in the model by appropriate statistical methods termed the parameterization.

Several types of models can be found in the literature from *zero-dimensional* (global-average) models to *general circulation* models of three dimensions. Each model type is suitable to answer a specific question. Thus, even a zero-dimensional model working with averages for the whole of the atmosphere (Earth) is sufficient to demonstrate the overall importance of atmospheric greenhouse gases, while simulations with general circulation models are necessary to describe changes in horizontal and vertical distribution of climatic variables (due to the increase of gas concentrations). General circulation models also take rather complicated feedback processes into consideration; however, highly efficient computer facilities are needed for the calculations. Table 6.3 summarizes the main characteristics of four general circulation models developed in

Table 6.3 Characteristics of Four General Circulation Models Used to Project Future Climatic Change[a]

Host Institution	Horizontal Resolution (lat. × long.)	Number of Layers in Vertical	Treat Diurnal Cycle	Model Mnemonic
National Center for Atmospheric Research (NCAR)	4.5° × 7.5°	9	No	CCM[b]
NOAA Geophysical Fluid Dynamics Laboratory (GFDL)	4.5° × 7.5°	9	No	GFDL[c]
NASA Goddars Institute for Space Studies (GISS)	7.8° × 10°	9	Yes	GISS[d]
Oregon State University (OSU)	4° × 5°	2	Yes	OSU[e]

Note: All models include realistic topography and geography, interactive and multiple-layer cloud cover, variable snow and sea-ice coverage, and interative soil moisture. Treatment of the oceans varies among models and for a given model with the type of simulation.

[a] MacCracken, 1990.

[b] Washington and Meehl, 1984.

[c] Manabe and Wetherald, 1987.

[d] Hansen et al., 1984.

[e] Schlesinger and Zhao, 1989.

the U.S. (MacCracken, 1990). In the table the horizontal and vertical resolutions of the models are given. The insufficient horizontal resolution imposes several problems if we are interested in potential regional climate changes. This is caused by the fact that gridpoints are the size of the state of Colorado, covering an area larger than most of the countries in Europe. In addition, it is believed that models produce reliable results on scales of several grid points with a typical dimension of 1000 km. This makes the evaluation of regional climate changes difficult, mainly if variable meteorological parameters like precipitation are considered. Finally, numerical studies indicate that the model representations are better in winter than in summer.

If we want to determine future climate variations by using computer models, before calculations we have to estimate possible changes in atmospheric composition, more exactly the modification of the concentration of atmospheric constituents should be forecast. This is possible by assuming potential scenarios for the anthropogenic emissions of different components.

6.4.2 Possible Emission Scenarios

If the greenhouse gas concentrations in the atmosphere increase, our climate system will reach a new equilibrium by warming in the troposphere. The warming is probable, in particular, if gases absorbing in the wavelength band from 7 to 13 μm are emitted, since in this band nearly 80% of the radiation emitted by the surface escapes to space (it is called an atmospheric "window") under undisturbed conditions. Unfortunately, most pollutant gases released by human activities absorb radiation in this band by "dirtying" the atmospheric window (Ramanathan, 1988). Future climate consequences of anthropogenic greenhouse gas emissions were recently estimated by the Intergovernmental Panel on Climate Change created jointly by the World Meteorological Organization and the United Nations Environment Program. The panel consisted of experts from all over the world under the leadership of B. Bolin (Sweden). The panel assumed four scenarios for greenhouse gas emissions (IPCC, 1990), which can be summarized as follows.

Scenario A, called the "Business as Usual" scenario, assumes that few or no steps are taken to limit greenhouse gas emissions. The energy supply is coal intensive and only a modest energy efficiency is achieved. Deforestation of tropical forests continues, and no control is introduced concerning agriculture methane and nitrous oxide release. Emissions of CFCs are reduced according to the Montreal Protocol (see Section 6.3.3) but only with a partial participation of the countries.

In Scenario B coal is significantly substituted by lower carbon fuels like natural gas. Energy is produced and used with a high efficiency. Deforestation is reversed and the Montreal Protocol accepted by all countries.

In Scenario C energy supply shifts to renewable and nuclear energy in the second half of the next century. CFCs are now phased out and agricultural emissions are limited.

Finally, in Scenario D the shift to renewable and nuclear energy is realized in the first half of the next century. In industrialized countries the emissions are controlled stringently with

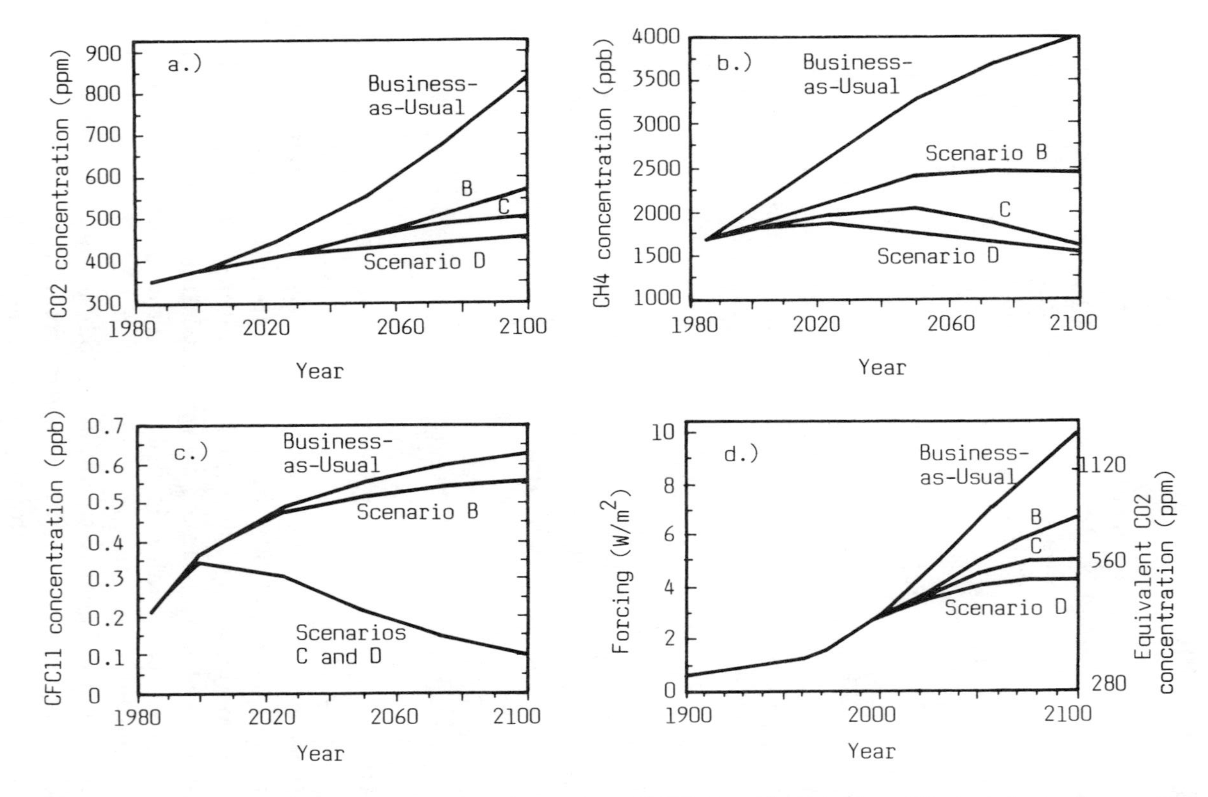

Figure 6.9 Atmospheric concentrations of carbon dioxide, methane, CFC-11, and equivalent CO_2 (see the text) according to IPCC (1990). In Figure 6.9d the climate forcing due to the increase of greenhouse gas concentrations is also plotted.

a moderate growth in developing countries. CO_2 emissions are reduced to 50% of 1985 levels by the middle of the next century.

Figure 6.9a-c shows the atmospheric concentrations of carbon dioxide, methane, and CFC-11 for these scenarios. It can be seen that in the case of Scenario A the CO_2 concentration in 2100 is estimated to be more than 800 ppm, while for Scenario D it is stabilized during the next century. Even Scenario B and C lead to a relatively modest CO_2 concentration increase by 2100. Methane reaches a very high level for the first scenario (4 ppm), while it is stabilized or begins to decrease in other cases. In agreement with our discussion in Section 6.3.3 the concentration of CFCs in the atmosphere decreases only when the phaseout is immediate and complete.

The effects of the concentration increase of different greenhouse gases can be determined by introducing the concept of *global warming potential*. This index defines the warming as a result of the release of a certain gas of unit mass/volume relative to the impact created by CO_2 emission of the same amount. By applying this concept the rise of the atmospheric level of greenhouse gases expressed in CO_2 concentration can be calculated as represented in Figure 6.9d. In this way the warming is determined for a given CO_2 concentration increase. In several model studies the climate response is calculated as double the carbon dioxide relative to the preindustrial level (280 ppm). For the scenarios A, B, and C outlined above this concentration will be reached in 2020, 2035, and 2045, respectively. In the case of the last scenario a total greenhouse gas doubling is foreseen for the end of the next century.

6.4.3 Result of Model Calculations

Climate models are validated by simulating past (see Chapter 1) and present climate variations. Over the past 100 years, when greenhouse gas concentrations have increased, the global temperature has raised by about 0.4 to 0.5°C (Figure 6.10), having been steadier in the Southern Hemisphere than in the Northern Hemisphere. The results of theoretical predictions are in agreement with this trend, but variations observed are around the lower limit expected. One can speculate that this partial discrepancy is caused by some factor (e.g., volcanic activity) that has moderated the warming. It can also be specu-

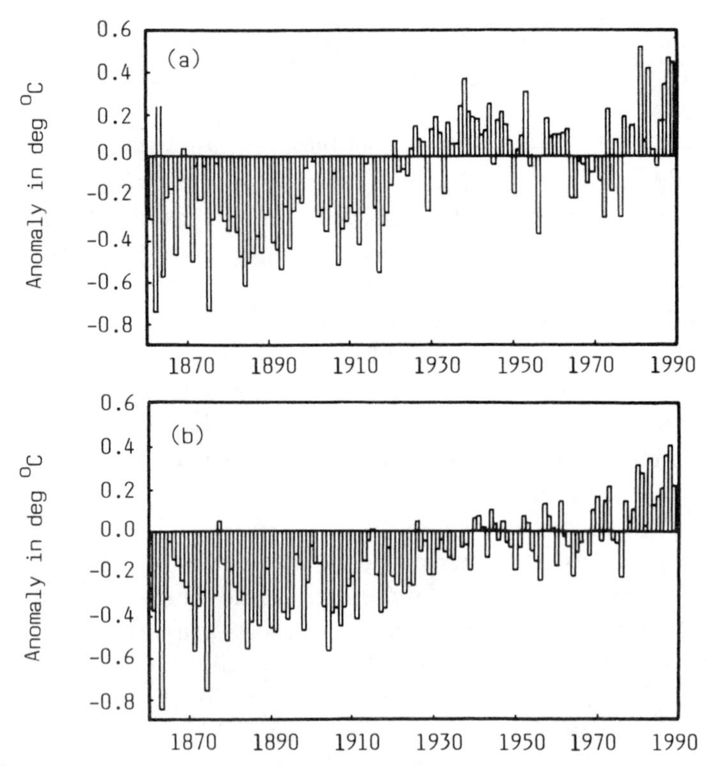

Figure 6.10 Land air temperatures, expressed as anomalies relative to 1951–1980 (data of Jones published in IPCC, 1990) (a) Northern Hemisphere, (b) Southern Hemisphere.

lated that climate response is not linear to radiative forcing* as assumed in models. Finally ,— and this is probably the most important (see Hasselmann, 1991) — it cannot be excluded that the oceans have slowed down the warming.

Calculations also show that 55% of the change in radiative forcing from 1980 to 1990 was produced by the increase of the CO_2 level. The relative contribution of CFCs 11-12 and methane was 17 and 15%, respectively. This means that carbon dioxide plays an important part in climate control because of its high concentration.

Concerning future temperature increase let us consider Figure 6.11a giving changes for the four scenarios in the last subsection. It can be seen that in the case of Scenario A ("Business as Usual") the increase in global mean temperature rela-

* Energy gain of the atmosphere due to the presence of greenhouse gases. The energy gain increases with increasing gas concentrations.

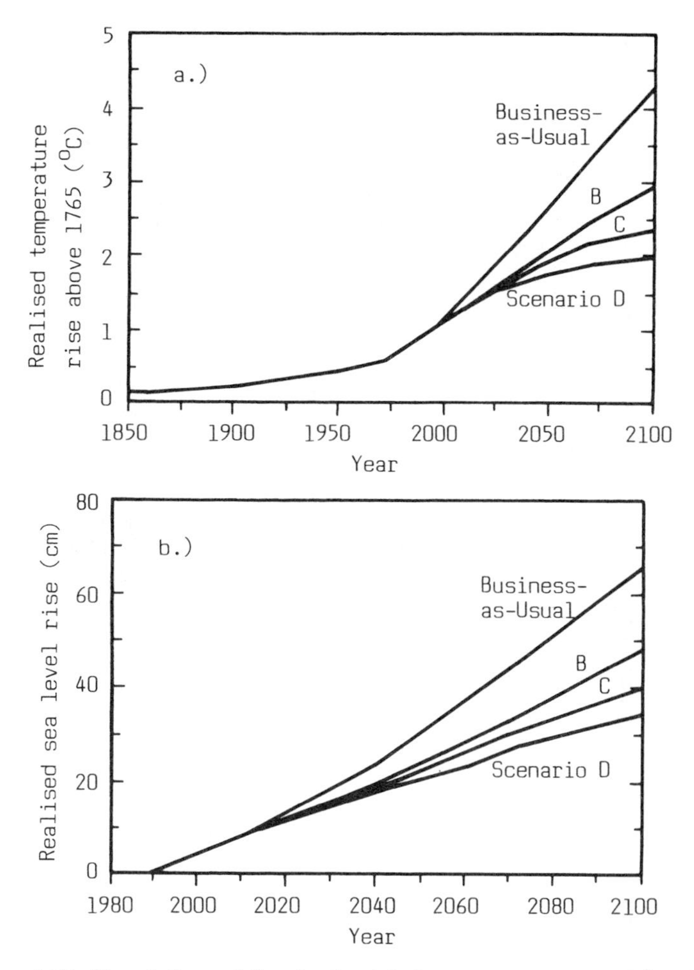

Figure 6.11 Simulations of the rise in global mean temperature (a) and sea level (b) for scenarios of IPCC (1990).

tive to 1765 will reach >4°C in 2100. Taking into account the high and low climate responses calculated, one can state that the estimate is between 3 and 6°C. Even the most favorable scenario (Scenario D) produces an average global warming of 2°C by 2100, which is still much higher than the magnitude of natural variations during the last centuries.

The variations of the warming are illustrated in Figure 6.12 for a doubling of CO_2 (Hansen et al., 1984). Figure 6.12a shows that warming in the troposphere is compensated by cooling in the stratosphere. Moreover, the warming in the surface air is significant near the poles, especially in the winter months (see also Figure 6.12b). This is due to the fact that warming reduces

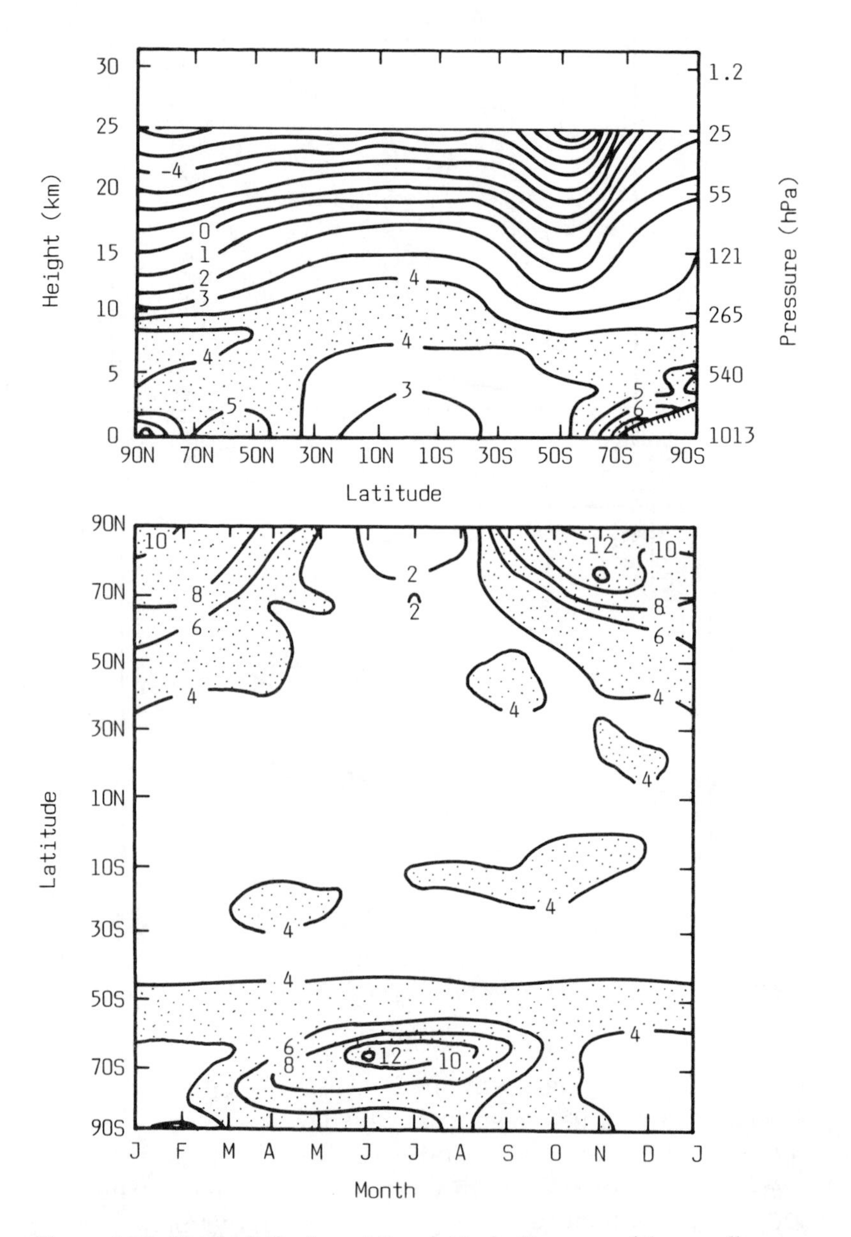

Figure 6.12 Height-latitude and time-latitude diagrams of the zonally averaged increase in surface temperature due to doubling CO_2 (IPCC, 1990). Warming >4°C stippled.

the extent of sea ice cover that decreases the surface albedo intensifying the warming process (a positive "temperature-albedo" feedback).

Further details of global warming are shown in Figure 6.13 where the global distribution of changes in surface air temperature due to CO_2 doubling are plotted. Figure 6.13a refers to winter months, while Figure 6.13b represents forecasting for June, July, and August. Again, the enhanced winter warming in higher latitudes is clearly shown, but the summer warming over the Arctic and Antarctic areas is somewhat smaller than the global mean. Further, in tropical regions the temperature increase is below the global mean and varies little with season. Finally, the increase of the summer temperature is relatively high over northern mid-latitude continents.

Changes in temperature obviously lead to changes in values of other climate parameters like evaporation and circulation, which is followed by variations of precipitation quantity. Results gained by general circulation models indicate that precipitation will be enhanced in high latitudes and the tropics during all seasons and in mid-latitude during wintertime. On the other hand, there is some indication that over continental areas in northern mid-latitude summer rainfall will decrease during the next century. This will influence significantly agricultural production, mainly if we take into account that soil moisture will also be reduced with the increase of evaporation. However, it should be noted that if the land surface becomes dryer than a critical level further evaporation is restricted, which may cause a reduction in the amount of low clouds. Reduction in evaporative cooling and cloud frequency will intensify the warming over the continents (see Figure 6.13).

6.4.4 Consequences of Climate Changes

It is evident that climate changes will produce important consequences for the human environment. Thus, it is estimated that the sea level will rise considerably in the future. This is supported by the curves of Figure 6.11b illustrating changes in sea level by 2100 for the IPCC (1990) scenarios. One can see that different scenarios predict a sea level rise between 35 and 65 cm. In the case of the "Business as Usual" scenario the average rate of global mean sea level rise is about 6 cm per decade over the next century with an uncertainty range of 3 to 10 cm. Of course, significant regional variations are possible beyond these global figures. Anyway, a sea level rise of such magnitude by the year 2100 could inundate a large part of present costal areas and could destroy many inland countries in the Pacific Ocean.

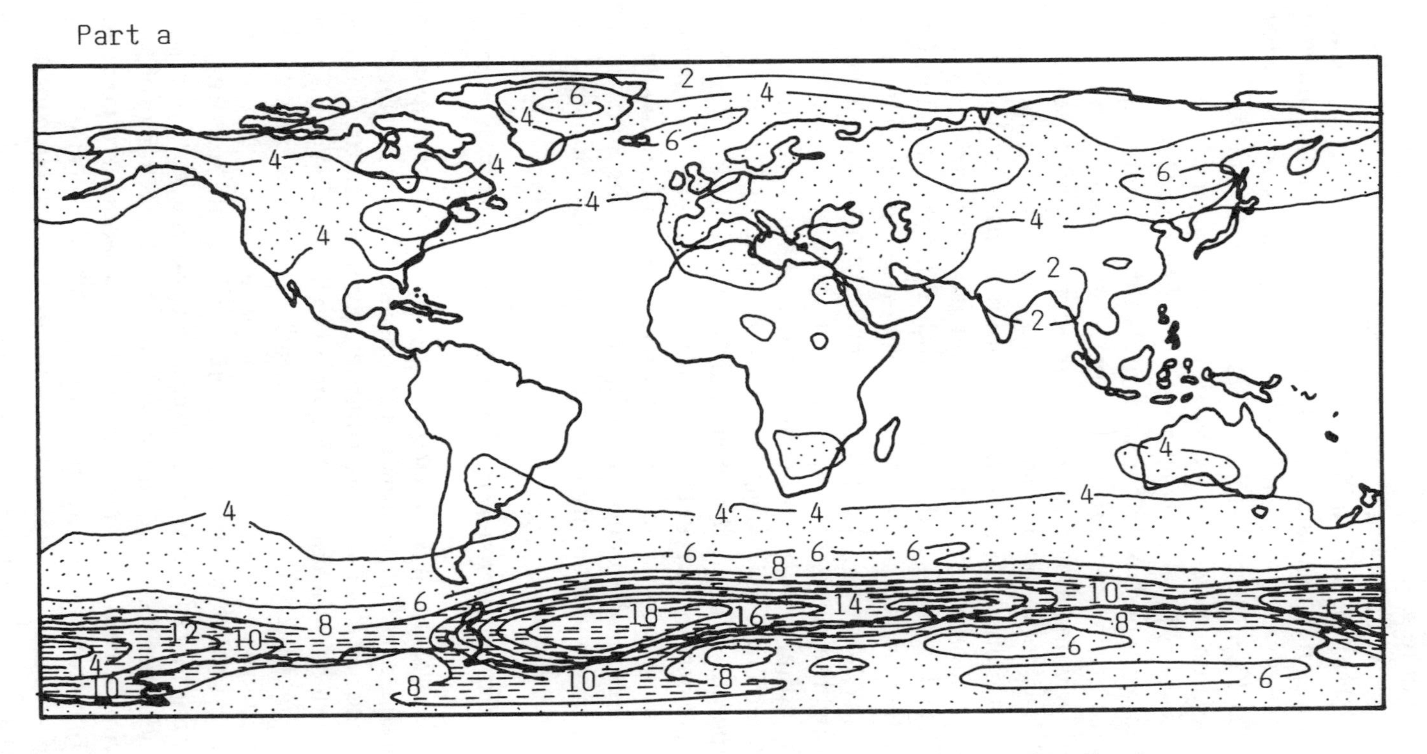

Figure 6.13 Change in surface air temperature (10 year mean) due to doubling CO_2 for winter (part a) and summer (part b) months as simulated by the GFDL model (see Table 6.3) (published in IPCC, 1990).

Figure 6.13 (continued)

Due to global warming life conditions of different *ecosystems* will also be modified, since they depend directly on local climate factors—temperature, precipitation, and soil moisture. The increase of carbon dioxide concentration alone may enhance the intensity of photosynthesis, but final effects can be reversed if the amount of water and other nutrients is limited. Further, respiration of plants is more sensitive to temperature than photosynthesis (Woodwell, 1989). A 1°C rise in temperature may increase the rate of respiration by 10 to 30%. Thus, global warming will speed the decay of organic matter without appreciable modifications of the rate of photosynthesis. This will lead to a positive feedback in CO_2 release and consequently in warming processes.

Since the activity of methanogenic bacteria is a direct function of the temperature one can postulate that CH_4 production will be increased by global warming. This is true in particular for high latitude ecosystems such as wet tundra and boreal wetlands. It is estimated (IPCC, 1990) that a 4°C rise in the temperature of soils in high latitudes could lead to a 45 to 60% increase of methane release from these systems. It can also be speculated that warming will intensify even N_2O release from the soil by accelerating the nitrogen cycle. All these speculations indicate that the emission of greenhouse gases will be accelerated by global warming.

In the middle latitudes a change of 1°C in mean temperature corresponds to a change in latitude of 100 to 150 km. This means that an important migration of the forest-prairie border is foreseen with a destruction of present forested areas. Forest destruction will further enhance the increase of CO_2 emission. Moreover, the position of present belts of the production of agricultural plants will be shifted. More exactly, areas convenient for producing a given plant will be displaced to higher latitudes or altitudes. This will certainly disturb the present-day food production activity of mankind.

Even this short discussion makes it evident that it would be of crucial importance to stabilize atmospheric composition and climate for preserving the present situation. However, important reductions in anthropogenic emissions would be necessary to obtain this goal. Thus, one can estimate that an emission reduction of more than 60% would be required to maintain CO_2 concentration at its present level.

The corresponding range for methane and nitrous oxide is 15 to 20 and 70 to 80%, respectively. At the same time, unfortunately, there is no international convention at all to mitigate the emission of greenhouse gases on a worldwide basis. Therefore, the conclusion of such a contract by countries of the world is urgently needed in spite of the economic and political problems involved.

6.5 IMPACTS OF INCREASING AEROSOL CONCENTRATION

6.5.1 Effects on Radiation Transfer and Clouds

It was mentioned in Chapter 1 that aerosol particles influence climate *directly* because they absorb and scatter solar radiation. Hence, anthropogenic modifications of the atmospheric aerosol burden can induce inadvertent changes in our climate system. The estimation of the effects of aerosol particles on climate is not an easy task (see Götz, 1991). Several workers tried to estimate aerosol effects by using climate models discussed previously. Thus, Coakly et al. (1983) calculated, by means of a one-dimensional (latitudinally dependent) energy balance model, that the global cooling of the Earth owing to the presence of tropospheric background aerosol is equal to 2 to 3 K. This conclusion was later principally proved by the simulations of Potter and Cess (1984) made by a two-dimensional approach. General circulation models were also used to determine possible effects of aerosol particles on climate. The studies of Coakly and Cess (1985) demonstrated that although aerosol impacts are sizable, they influence only slightly the temperatures calculated by the model. Significant changes in temperature were calculated only for isolated portions of the world (e.g., over a region of Africa) where desertification is significant.

We have seen in Section 3.4.2 that a major part of fine aerosol particles in the troposphere consists of sulfates. Thus, the estimation of the effects of sulfate particles formed from anthropogenic sulfur dioxide is of crucial importance. The effects of sulfate particles on the climate were first estimated by Bolin and Charlson (1976) for the air over the eastern U.S. and Western Europe. According to their results the reduction of solar radiation caused by the scattering (H_2SO_4 and its ammonium salts do not absorb solar radiation) of sulfur particles is between 5 and 10% on a regional scale. Such a reduction induces a temperature decrease of several degrees. This approach was further developed by Charlson et al. (1991) who calculated the global distribution of the column mass concentration of sulfate particles by a three-dimensional budget model. On the basis of column concentration and other information, these authors determined the optical depth* of the tropospheric sulfate and, finally, the scattering caused by the particles under clean sky conditions. They concluded that the change of reflected solar flux due to anthropogenic sulfate particles over the Northern Hemisphere is somewhat higher than -1 W m^{-2}. Presently the total radiative forcing of greenhouse gases (see Figure 6.9d) is around 1.5 W m^{-2}, which is only slightly higher than the value received by Charlson and his co-workers for man-made sulfate aerosols. It was also found that aerosol effects are significant in particular over the industrialized regions of the continents in the Northern Hemisphere.

Many experts believe that the *indirect* climate effects of aerosol particles are even more important than direct impacts.** This is due to the fact that a fraction of aerosol particles serves as cloud condensation nuclei (CCN), the concentration of which influences cloud structure and lifetime (see Section 4.2.1). If the number of CCN increases (e.g., from anthropogenic SO_2 emissions) clouds becomes more stable, since the cloud droplet concentration increases and the average droplet size decreases. Such clouds have a higher albedo value than other clouds, which means that they scatter very efficiently incoming solar radiation (Twomey, 1977).

It was mentioned previously that global warming will lead to the intensification of evaporation. In climate model studies discussed in Section 6.4 one assumes that water vapor remains in the atmosphere, except above a certain relative humidity,

---

* Reduction (extinction) of solar radiation in a given layer of the atmosphere.

** This opinion does not hold for such extreme situations as a large-scale nuclear war (Schneider and Thompson, 1988) or oil fires after the Gulf War (Browning et al., 1991). Such extreme cases are not discussed in this volume, hoping that they will not occur in the future.

and contributes to the magnitude of global warming, since H_2O is a very effective greenhouse component (Ramanathan, 1988; Twomey, 1991). Moreover, in usual climate models a parameterization procedure is used to consider condensation. This means that above a certain relative humidity (e.g., 99 to 100%) moisture excess is numerically precipitated. However, the structure of clouds (i.e., CCN effects) formed is not taken into account in further calculations. In other words, the indirect impacts of increasing aerosol concentration on climate through the modification of cloud structure are ignored. Generally speaking, it is an essential issue whether water due to enhanced evaporation remains in vapor phase or stays in the atmosphere in clouds. In the first case water increases, in the second it decreases the surface temperature. Thus, the proper simulation of water cycle in climate models would be crucial.

The effects of increasing aerosol concentration on clouds and consequently on climate were estimated by Charlson et al. (1987) by a relatively simple model calculation. They found that an increase of 30% in CCN/cloud droplet concentration in stratiform clouds over the ocean causes a decrease of 10% in average droplet size. Above these marine clouds such a change in cloud structure induces a planetary albedo* increase of 0.016 over the oceans, which is equivalent to a value of 0.005 averaged over the entire Earth. It can be calculated that such an albedo increase leads to a temperature decrease of 1.3 K in the surface air. This value is about a third of the average global warming predicted for a doubling of atmospheric CO_2 (see previous section).

However, we must take into account the fact that clouds also absorb infrared radiation emitted by the Earth's surface and that this can contribute to greenhouse effects. Model calculations show (Grassl, 1982) that the energy loss due to the increase of the cloud albedo in the shortwave band is not compensated by a simultaneous reduction of emission in the longwave range. This theoretical conclusion is supported by satellite observations, indicating that clouds have a net cooling effect on the Earth (Ramanathan et al., 1989). One can also consider that insoluble aerosol particles, like elemental carbon (soot) partly of anthropogenic origin, can also be imbedded

* Note that the actual value is 0.3 (see Chapter 1).

into cloud droplets because of the conversion of sulfur dioxide to water-soluble sulfate on their surface (see Chapter 3). Elemental carbon particles covered by a solution layer can act as CCN. The "dirty" cloud droplets formed on these particles reduce solar radiation intensity very efficiently (Chylek et al., 1984; Kondratyev and Binenko, 1987), which leads to a further decrease in surface air temperature.

If the above discussion is correct, the reflectivity of clouds over the Northern Hemisphere should be different from cloud reflectivity in the Southern Hemisphere. That is, clouds in the Northern Hemisphere form on more numerous CCN owing primarily to anthropogenic SO_2 emissions. If we suppose a CCN concentration ratio of 3:1 between the two hemispheres, the calculations result in an albedo difference of 0.023 (Schwartz, 1988). However, this is probably an overestimate since the distribution of sulfate particles is not uniform in the atmosphere over the Northern Hemisphere (see Figure 3.6). Even by considering this possibility Schwartz concluded that the interhemispheric albedo difference is not smaller than 0.008, with the greater value in the Northern Hemisphere. Keeping this value in mind Schwartz compared total albedos and their cloud and clear-sky components measured over the two hemispheres as a function of latitudes. Since no differences were found, he concluded that the control of albedo and temperature is too complex to be governed by a single variable like sulfate concentration. The problem was later reconsidered by Wigley (1989) by comparing observed differences in hemispheric mean temperatures and figures calculated by a simple climate model. He found that during the last decades the Northern Hemisphere has been sensibly cooler than the Southern Hemisphere; this agrees with model calculations taking into account the effects of increasing sulfate particle concentration. On this basis Wigley postulated that the effects of anthropogenic SO_2 emissions may have compensated the temperature changes that have resulted from the greenhouse effect.

There is no intention here to close this subject with a final conclusion. However, although numerical figures discussed are not certain in an absolute sense, it is probable that global warming caused by the concentration rise of greenhouse gases in the atmosphere will be moderated by the direct and indirect effects of increasing aerosol concentration. Thus, the more precise determination of these impacts by appropriate climate modeling is a challenge for the scientific community in the years to come.

6.5.2 Impacts on Air Quality of Arctic Regions

Polar areas play an important part in climate regulation, since the extension of ice cover determines in a great measure the albedo of the Earth's surface. Changes in the extent of polar ice cover (e.g., owing to temperature variations) are followed by albedo changes (the albedo of ice is much higher than that of other surfaces) which intensify the initial process (positive feedback). On the other hand, any change in temperature difference between the poles and the Equator may lead to a loss in the driving force for the atmospheric and oceanic "heat engine" controlling the general circulation.

This is particularly important for our subject. Due to the tropospheric circulation system (Raatz, 1985), a significant part of secondary pollutants, formed from precursors emitted to the atmosphere over the Northern Hemisphere, is transported to areas over the North Pole, where they are accumulated to produce a phenomenon called the *Arctic haze.* The Arctic haze is observed during late winter and early spring, when wet removal processes in this area are relatively inefficient. It consists of an aerosol layer in which the total number concentration is typically 500 to 3000 cm^{-3} (Herbert et al., 1987). Figure 6.14 represents the situation according to airborne measurements carried out on April 2–3, 1986. It can be seen that between 900 and 800 hPa above the temperature inversion the number concentration approaches 10,000 cm^{-3}. In contrast, in summer the concentrations measured in Arctic regions (Jaenicke and Schütz, 1982) are rather similar to those observed in Antarctic areas (50 to 100 cm^{-3}, see also Section 3.4).

Although the particles constituting the Arctic haze are composed partly of sulfate species (Rahn, 1981) elemental carbon essentially controls the composition (Rosen et al., 1981). Since carbon is a very effective absorber of solar radiation, it is suggested (e.g., Chylek et al., 1984) that Arctic haze will warm the air over the polar regions, and after deposition soot carbon will decrease the albedo of polar snow cover. Rosen et al. (1984) speculate that atmospheric heating due to Arctic aerosol

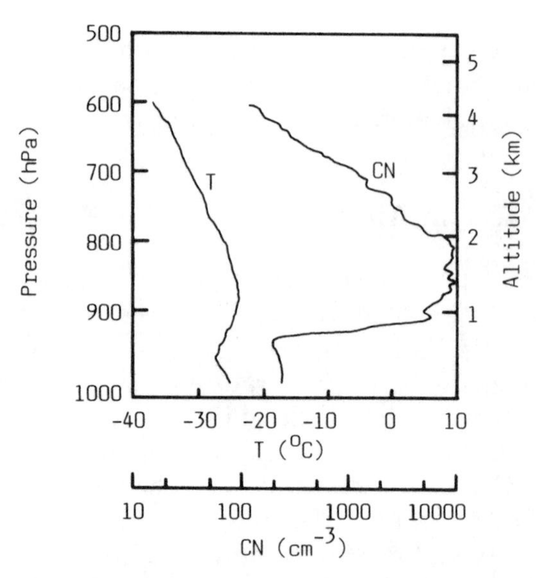

Figure 6.14 Vertical profiles of total aerosol number (CN) and temperature (T) in an Arctic haze layer (Herbert et al., 1987).

is nearly the same as a doubling of the concentration of CO_2. Thus, the accumulation of man-made aerosol particles in the atmosphere over the Arctic further complicates the evaluation of large-scale climatic consequences of human activities.

6.6 CLOSING REMARKS

6.6.1 The Sustainable Development

Philosophically speaking, all substances emitted by human activities to the atmosphere are harmful in a certain way, since they disturb some natural process. The materials having a long residence time like most of the greenhouse gases are mixed in the entire troposphere to cause such worldwide problems as global warming. In addition, pollutant gases with long lifetimes reach the stratosphere to modify stratospheric ozone chemistry. Thus, such compounds as chlorofluorocarbons, found to be absolutely harmless for human life in the surface air, produce dangerous effects in higher levels of the atmosphere by altering the state of the ozone layer, which protects the biosphere from lethal radiations. On the other hand, reactive anthropogenic species like sulfur and nitrogen oxides leave the atmosphere relatively quick to be deposited onto other media of our environment. This is certainly a benefit for the atmosphere, but excess input of any materials into soils, lakes, and surface waters can cause serious regional impacts such as acidification. Even such a gas as ammonia, which neutralizes acid aerosol particles and droplets in the air, intensifies acidification problems in the soil.

Chemical reactions of man-made species generally create more dangerous secondary products, which modify, among other things, the oxidizing capacity of the air and consequently the cycles of other substances in nature. By changing atmospheric composition on global and regional scales, mankind is conducting a huge experiment resulting in unintended consequences which will alter conditions necessary for human life. One can speculate that, besides the possibility of a nuclear war, the modification of atmospheric composition by different pollutants and their impacts poses the greatest threat to the future of human society. Hence, it is of crucial importance to protect the atmosphere from pollution as far as possible.

This is a great challenge for the human race since the growing population, mostly in the developing part of the world, needs more energy, more industrial production, and last but not least, more food. On the other hand, people living in the developed part of the world believe that the tolerance of the environment is infinite, and they consume ten times more energy per capita than people living in other countries. The great challenge is to elaborate the manners of further economic development that at the same time, do not destroy the environment or jeopardize the welfare of coming generations. This so-called *sustainable development* is technically not utopian if a new model of wealth is introduced—one that is "characterized by low pollution, low resource and energy consumption and low land destruction at equal or better living quality" (Weizsäcker, 1990). Realizing this depends upon reforming the present distribution of resources and economy, including prices and tax systems.

At the same time the scientific community should understand in a more proper way, the mechanism of atmospheric operation. For obtaining this goal new monitoring systems are needed, like the Global Atmosphere Watch of the World Meteorological Organization, to provide data and other information on the chemical composition and related physical characteristics of our atmosphere. Moreover, research efforts must be coordinated as planned in the International Geosphere-Biosphere Programme (IGBP) of ICSU* aiming to understand present and future global changes in our environment. The core project of IGBP, the International Global Atmospheric Chemistry Programme is particularly important for our subject, since its goal is to investigate atmospheric composition and interactions between human activities and chemical processes. Finally, human dimensions and economic aspects of environmental changes should also be considered, and scientists must formulate their results in a way understandable for the policy makers and general public.

6.6.2 The Problem of Biospheric Control

In Chapter 1 it was mentioned that the biosphere can potentially control the environment, including the atmosphere. Consequently, the possibility of a negative biospheric feedback to anthropogenic changes cannot be entirely ruled out. This view is based on the fact that in the past physical and chemical conditions in the environment have always been favorable for life. On this basis Lovelock and Margulis (1974) postulate a theory, called the *Gaia hypothesis,* according to which the biosphere intentionally creates an optimum environment for itself. The Gaia hypothesis claims that the composition of the atmosphere, like in a living organism, is far from physicochemical equilibrium and each atmospheric component has its own special purpose; the atmosphere is a part of the biosphere. If this idea is correct one can speculate that the Earth is a self-regulating system that controls the chemical composition and climate at a steady state optimal for life. Obviously it would be difficult to formulate the Gaia hypothesis in a form suitable for scientific control. To overcome this difficulty Watson and Lovelock (1983) invented an artificial planet, called Daisyworld. In this world two types of daisies (black and white) with different albedos absorb the incoming radiation to a different degree. By using a model borrowed from the population ecology, Watson and Lovelock demonstrate that, if external conditions vary, the presence of the

* International Council of Scientific Unions.

biosphere stabilizes the temperature because of the effects of changes in the extension of the areas covered by black and white daisies on the albedo of the planet. The same conclusions were obtained by Mészáros and Pálvölgyi (1990) who imagined the planet with an atmosphere consisting of a greenhouse gas and an aerosol layer. Their results show that the climatic impacts of changes in greenhouse gas amount are significantly moderated by the presence of daisies. However, Zeng et al. (1990) found that periodic, and even chaotic, states on Daisyworld can exist for certain values of the parameter controlling the feedback between biota and temperature.

There is no intention here to go further into the details of this interesting debate. It is sufficient for our purpose to state that the potential self-regulating capacity of Nature guarantees only the conditions for life as a whole and not for a certain species of the biosphere like man. In other words, anthropogenic changes in atmospheric composition do not endanger the entire biosphere, but they can destroy the present situation favorable for a species, man. Thus, even if some Gaia-like feedback exists, we have to protect our atmospheric environment in our own interest. This is absolutely necessary since, as discussions in this volume illustrate, our atmosphere, this fragile extraordinary gas mixture, is the consequence of life on Earth. Its composition is a measure of the state of this living planet. Therefore, we can conclude with Lovelock (1988) that "The atmosphere is the face of the planet, and it tells, just as do our faces, its state of health and even if it is alive or dead." We have to do everything to preserve the health of our unique planet.

References

Alcamo, J., Bartnicki, J., and Schöpp, W. 1990. Long-range transport of sulfur and nitrogen compounds in Europe's atmosphere, in *The RAINS Model of Acidification. Science and Strategies in Europe,* J. Alcamo, R. Shaw, and L. Hordijk, Eds., Kluwer Academic Publishers, Dordrecht, Netherlands, 115–178.

Amann, M. 1990. Energy use, emissions, and abatement costs, in *The RAINS Model of Acidification. Science and Strategies in Europe,* J. Alcamo, R. Shaw, and L. Hordijk, Eds., Kluwer Academic Publishers, Dordrecht, Netherlands, 61–114.

Andreae, M.O. 1986. The oceans as a source of atmospheric sulfur, in *The Role of Air-Sea Exchange in Geochemical Cycling,* P. Buat-Ménard, Ed., D. Reidel Publishing Co., Dordrecht, Netherlands, 331–362.

Andreae, M.O., Charlson, R.J., Bruynseels, F., Storms, H., Van Grieken, R., and Maenhaut, W. 1986. Internal mixture of sea salt, silicates, and excess sulfate in marine aerosols. *Science* 232:1620–1623.

Andreae, M.O., Talbot, R.W., Andreae, T.W., and Harriss, R.C. 1988. Formic and acetic acid over the central Amazon Region, Brazil, 1, dry season. *J. Geophys. Res.* 93:1616–1624.

Ashmore, M., Bell, N., and Rutter, J. 1985. The role of ozone in forest damage in West Germany. *Ambio* 14:81–87.

Asman, W. and Janssen, A. 1987. A long-range transport model for ammonia and ammonium for Europe. *Atmos. Environ.* 21:2099–2119.

Atkinson, R.J., Matthews, W.A., Newman, P.A., and Plumb, R.A. 1989. Evidence of the mid-latitude impact of Antarctic ozone depletion. *Nature* 340:290–294.

Barett, J.W., Solomon, P.M., de Zafra, R.L., Jaramillo, M., Emmons, L., and Parrish, A. 1988. Formation of the Antarctic ozone hole by the ClO dimer mechanism. *Nature* 336:455–458.

Barrie, L.A. and Hales, J.M. 1984. The spatial distributions of precipitation acidity and major ion wet deposition in North America during 1980. *Tellus* 36B:333–355.

Barrie, L.A. and Schemenauer, R.S. 1989. Wet deposition of heavy metals, in *Control and Fate of Atmospheric Trace Metals,* J.M. Pacyna and B. Ottar, Eds., Kluwer Academic Publishers, Dordrecht, Netherlands, 203–231.

Bates, T.S., Cline, J.D., Gammon, R.H., and Kelly-Hansen, S.R. 1987. Regional and seasonal variations in the flux of oceanic dimethylsulfide to the atmosphere. *J. Geophys. Res.* 92:2930–2938.

Beilke, S. 1985. *Acid Deposition.* Umweltbundesamt, Frankfurt.

Beilke, S. and Gravenhorst, G. 1987. Deposition, in *Evaluation of Atmospheric Processes Leading to Acid Deposition in Europe,* Air Pollution Research Report 10, Commission of the European Communities, 113–135.

Bekki, S., Toumi, R., Pyle, J.A., and Jones, A.E. 1991. Future aircraft and global ozone. *Nature* 354:193–194.

Berner, R.A. 1990. Atmospheric carbon dioxide levels over phanerozoic time. *Science* 249:1382–1386.

Berresheim, H. and Jaeschke, W. 1983. The contribution of volcanoes to the global atmospheric sulfur budget. *J. Geophys. Res.* 88:3732–3740.

Berresheim, H., Andreae, M.O., Iverson, R.L., and Li, S.M. 1991. Seasonal variations of dimethylsulfide emissions and atmospheric sulfur and nitrogen species over the western North Atlantic Ocean. *Tellus* 43B:353–372.

Bigg, E.K. 1986. Ammonium compounds in stratospheric aerosols. *Tellus* 38B:62–66.

Bigg, E.K., Gras, J.L., and Evans, C. 1984. Origin of Aitken particles in remote regions of the Southern Hemisphere. *J. Atmos. Chem.* 1:203–214.

Bingemer, H.G. and Crutzen, P.J. 1987. Production of methane from solid wastes. *J. Geophys. Res.* 92:2181–2187.

Blanchard, D.C. 1969. The oceanic production rate of cloud nuclei. *J. Rech. Atmos.* 4:1–6.

Blanchard, D.C., Woodcock, A.H., and Cipriano, R.J. 1984. The vertical distribution of the concentration of sea salt in the marine atmosphere near Hawaii. *Tellus* 36B:118–125.

Blifford, J.H. 1970. Tropospheric aerosols. *J. Geophys. Res.* 75:3099–3103.

Boden, T.A., Kanciruk, K., and Farrel, M.P. 1990. *TRENDS '90. A Compendium of Data on Global Change,* R.J. Sepanski and F.W. Stoss, Eds., Carbon Dioxide Information Analysis Center, Oak Ridge, Tennessee.

Bojkov, R., Bishop, L., Hill, W.J., Reinsel, G.C., and Tiao, G.C. 1990. A statistical trend analysis of revised Dobson total ozone data over the northern hemisphere. *J. Geophys. Res.* 95:9785–9808.

Bolin, B. and Charlson, R.J. 1976. On the role of the tropospheric sulfur cycle in the shortwave radiative climate of the Earth. *Ambio* 5:47–54.

Bónis, K., Mészáros, E., and Putsay, M. 1980. On the atmospheric budget of nitrogen compounds over Europe. *Idöjárás* 84:57–67.

Borbély-Kiss, I., Bozó, L., Koltay, E., Mészáros, E., Molnár, A., and Szabó, Gy. 1991. Elemental composition of aerosol particles under background conditions in Hungary. *Atmos. Environ.* 25A:661–668.

Böttger, A., Ehhalt, D.H., and Gravenhorst, G. 1978. Atmospharische Kreislaufe von Stickoxiden und Ammoniak. *Ber. Kernforschungsanlage Jülich*, Nr. 1558.

Boutron, C.F., Patterson, C.C., Petrov, V.N., and Barkov, N.I. 1987. Preliminary data on changes of lead concentrations in Antarctic ice from 155,000 to 26,000 years BP. *Atmos. Environ.* 21:1197–1202.

Braaten, D.A. and Cahill, T.A. 1986. Size and composition of Asian dust transported to Hawaii. *Atmos. Environ.* 20:1105–1109.

Brasseur, G.P. 1991. A deepening, broadening trend. *Nature* 352:668–669.

Brasseur, G.P., Granier, C., and Walters, S. 1990. Future changes in stratospheric ozone and the role of heterogeneous chemistry. *Nature* 348:626–628.

Breitenbeck, G.A., Blackmer, A.M., and Bremner, J.M. 1980. Effects of different nitrogen fertilizers on emission of nitrous oxide from soil. *Geophys. Res. Lett.* 7:85–87.

Bremner, J.M., Robbins, S.G. and Blackmer, A.M. 1980. Seasonal variability in emissions of nitrous oxide from soil. *Geophys. Res. Lett.* 7:641–644.

Bremner, J.M. and Blackmer, A.M. 1981. Terrestrial nitrification as a source of atmospheric nitrous oxide, in *Denitrification, Nitrification and Atmospheric Nitrous Oxide,* C.C. Delwiche, Ed., John Wiley & Sons, Inc., New York, 151–170.

Brimblecombe, P. and Pitman, J. 1980. Long-term deposit at Rothamsted, southern England. *Tellus* 32:261–267.

Browning, K.A., Allam, R.J., Ballard, S.P., Barnes, R.T.H., Bennetts, D.A., Maryon, R.H., Mason, P.J., McKenna, D., Mitchell, J.F.B., Senior, C.A., Slingo, A., and Smith, F.B. 1991. Environmental effects from burning oil wells in Kuwait. *Nature* 351:363–367.

Brühl, C. and Crutzen, P. 1989. On the disproportional role of tropospheric ozone as a filter against solar UV-B radiation. *Geophys. Res. Lett.* 16:703–706.

Brune, W.H., Toohey, D.W., and Anderson, J.G. 1990. *In situ* observations of ClO in the Arctic stratosphere: ER-2 aircraft results from 59°N to 80°N latitude. *Geophys. Res. Lett.* 17:505–508.

Buijsman, E. and Erisman, J.-W. 1988. Wet deposition of ammonium in Europe. *J. Atmos. Chem.* 6:265–280.

Byers, H.R., Sievers, J.R., and Tufts, B.J. 1957. Distribution in the atmosphere of certain particles capable of serving as condensation nuclei, in *Artificial Stimulation of Rain*, H. Weickmann and W. Smith, Eds. Pergamon Press, New York, 47–70.

Cachier, H., Buat-Ménard, P., Fontugne, M., and Chesselet, R. 1986. Long-range transport of continentally-derived particulate carbon in the marine atmosphere: evidence from stable carbon isotope studies. *Tellus* 38B:161–177.

Chamenides, W.L., Stedman, D.H., Dickerson, R.R., Rusch, D.W., and Cicerone, R.J. 1977. NO_x production in lightning. *J. Atmos. Sci.* 34:143–149.

Chang, T.Y. 1984. Rain and snow scavenging of HNO_3 vapor in the atmosphere. *Atmos. Environ.* 18:191–197.

Chapellaz, J., Barnola, J.M., Raynaud, D., Korotkevich, Y.S., and Lorius, C. 1990. Ice-core record of atmospheric methane over the past 160,000 years. *Nature* 345:127–131.

Chapman, S. 1930. On ozone and atomic oxygene in the upper atmosphere. *Phil. Mag.* 10:369–383.

Charlson, R.J. and Rodhe, H. 1982. Factors controlling the acidity of natural rainwater. *Nature* 295:683–685.

Charlson, R.J., Lovelock, J.E., Andreae, M.O., and Warren, S.G. 1987. Oceanic phytoplankton, atmospheric sulfur, cloud albedo and climate. *Nature* 326:655–661.

Charlson, R.J., Langner, J., Rodhe, H., Leovy, C.B., and Warren, S.G. 1991. Perturbation of the northern hemisphere radiative balance by backscattering from anthropogenic sulfate aerosols. *Tellus* 43A-B:152–163.

Chylek, P., Ramaswamy, V., and Srivastava, V. 1984. Graphitic carbon content of aerosols, clouds and snow, and its climatic implications. *Sci. Total Environ.* 36:117–120.

Cicerone, R. and Oremland, R. 1988. Biogeochemical aspects of atmospheric methane. *Global Biogeochem. Cycles* 2:299–327.

Cicerone, R.J., Elliott, S., and Turco, R.P. 1991. Reduced Antarctic ozone depletions in a model with hydrocarbon injections. *Science* 254:1191–1191.

Coakly, J.A., Cess, R.D., and Yurevich, F.B. 1983. The effect of tropospheric aerosols on the Earth's radiation budget: a parametrization for climate models. *J. Atmos. Sci.* 40:116–138.

Coakly, J.A. and Cess, R.D. 1985. Response of the NCAR Community Climate Model to radiative forcing by the naturally occurring tropospheric aerosol. *J. Atmos. Sci.* 42:1677–1692.

Cohen, Y. and Gordon, L.I. 1979. Nitrous oxide production in the ocean. *J. Geophys. Res.* 84:347–353.

CONCAWE. 1986. Volatile organic compound emissions: an inventory for Western Europe. Report No. 2/86, Den Haag.

Crutzen, P.J. 1971. Ozone production rates in an oxygen-hydrogen-nitrogen oxide atmosphere. *J. Geophys. Res.* 76:7311–7327.

Crutzen, P.J. 1974. Photochemical reactions by and influencing ozone in the troposphere. *Tellus* 26:47–57.

Crutzen, P.J. 1976. The possible importance of CSO for the sulfate layer of the stratosphere. *Geophys. Res. Lett.* 3:73–76.

Crutzen, P.J. 1982. The global distribution of hydroxyl, in *Atmospheric Chemistry*, E.D. Goldberg, Ed., Dahlem Konferenzen: Springer-Verlag, 313–328.

Crutzen, P.J. and Brühl, C. 1989. The impact of observed changes in atmospheric composition on global atmospheric chemistry and climate, in *The Environmental Record in Glaciers and Ice Sheets*, H. Oeschger and C.C. Langway, Jr., Eds., John Wiley & Sons, New York, 249–266.

Crutzen, P.J. and Andreae, M.O. 1990. Biomass burning in the tropics: impact on atmospheric chemistry and biogeochemical cycles. *Science* 250:1669–1678.

Crutzen, P.J. and Zimmermann, P.H. 1991. The changing photochemistry of the troposphere. *Tellus* 43A-B:136–151.

Cullis, C.F. and Hirschler, M.M. 1980. Atmospheric sulfur: natural and man-made sources. *Atmos. Environ.* 14:1263–1278.

Davies, T.D. and Nicholson, K.W. 1982. Dry deposition velocities of aerosol sulfate in rural eastern England, in *Deposition of Atmospheric Pollutants*, H.W. Georgii and J. Pankrath, Eds., D. Reidel Publishing Co., Dordrecht, Netherlands, 31–42.

Day, J.A. 1963. Small droplets from rupturing air-bubble films. *J. Rech. Atmos.* 1:191–196.

De Leeuw, F.A.A.M. and Van Rheineck Leyssius, H.J. 1991. Calculation of long-term averaged and episodic oxidant concentrations for The Netherlands. *Atmos. Environ.* 25A:1809–1818.

Delwiche, C.C. 1978. Biological production and utilization of N_2O. *Pure Appl. Geophys.* 116:414–422.

Derwent, R.G. 1986. *The Nitrogen Budget for the United Kingdom and Northwest Europe.* ETSU Report 37. ETSU, Harwell, Oxfordshire, UK.

Detwiler, R.P. and Hall, C.A.S. 1988. Tropical forests and the global carbon cycle. *Science* 239:42–47.

Dignon, J. and Hameed, S. 1989. Global emissions of nitrogen and sulfur oxides from 1860 to 1980. *J. Air Pollut. Control Assoc.* 39:180–186.

Duce, R.A. 1978. Speculations on the budget of particulate and vapor phase non-methane organic carbon in the global troposphere. *Pure Appl. Geophys.* 116:244–248.

Duce, R.A. 1986. *Air-sea Interchange of Pollutants.* WMO Environmental Pollution Monitoring and Research Programme, No. 37, Geneva.

Ehhalt, D.H., Rudolph, J., and Schmidt, U. 1986. On the importance of light hydrocarbons in multiphase atmospheric systems, in *Chemistry of Multiphase Atmospheric Systems,* W. Jaeschke, Ed., NATO ASI Series, Vol. G6, Springer-Verlag, Berlin, 321–350.

Eliassen, A. 1978. The OECD study of long range transport of air pollutants: long range transport modelling. *Atmos. Environ.* 12:479–487.

Farlow, N.H., Oberbeck, V.R., Snetsinger, K.G., Ferry, G.V., Polkowski, G., and Hays, D.M. 1981. Size distributions and mineralogy of ash particles in the stratosphere from eruptions of Mt. St. Helens. *Science* 211:832–834.

Farman, J.C., Gardiner, B.G., and Shanklin, J.D. 1985. Large losses of total ozone in Antartica reveal seasonal ClO_x/NO_x interaction. *Nature* 315:207–210.

Fehsenfeld, F. 1990. Global inventories of terrestrial biogenic emissions. Lecture presented at the Conference on Chemistry of the Global Atmosphere. Chamrousse, France.

Finlayson-Pitts, B.J. and Pitts, J.N., Jr. 1986. *Atmospheric Chemistry: Fundamentals and Experimental Techniques.* John Wiley & Sons, New York.

Fishman, J. and Crutzen, P.J. 1978. The origin of ozone in the troposphere. *Nature* 274:855–858.

Fishman, J., Fakhruzzaman, K., Cros, B., and Nganga, D. 1991. Identification of widespread pollution in the southern hemisphere deduced from satellite analyses. *Science* 252:1693–1696.

Frakes, L.A. 1979. *Climates Throughout Geologic Time.* Elsevier, New York.

Friend, H.P. 1973. The global sulfur cycle, in *Chemistry of the Lower Atmosphere,* S.I. Rasool, Ed., Plenum Press, New York, 177–201.

Galloway, J.N. and Whelpdale, D.M. 1980. An atmospheric sulfur budget for eastern North America. *Atmos. Environ.* 14:409–417.

Galloway, J.N., Thornton, J.D., Norton, S.A., Volchok, H.L., and McLean, R.A.N. 1982. Trace metals in the atmospheric deposition: a review and assessment. *Atmos. Environ.* 16:1677–1700.

Galloway, J.N. and Rodhe, H. 1991. Regional atmospheric budgets of S and N fluxes: how well can they be quantified? *Proc. R. Soc. Edinburgh* 97B:61–80.

Garland, J.A. 1978. Dry and wet removal of sulfur from the atmosphere. *Atmos. Environ.* 12:349–362.

Georgii, H.-W. 1982. *Review of the chemical composition of precipitation as measured by the WMO BAPMon.* WMO Environmental Pollution Monitoring Programme.

Götz, G. 1991. Aerosols and climate, in *Atmospheric Particles and Nuclei,* G. Götz, E. Mészáros, and G. Vali, Eds., Akadémiai Kiadó, Budapest, 193–242.

Grassl, H. 1982. The influence of aerosol particles on radiation parameters of clouds. *Idöjárás* 86:60–75.

Greenberg, J.P., Zimmerman, P.R., Heidt, L., and Pollock, W. 1984. Hydrocarbon and carbon monoxide emissions from biomass burning in Brazil. *J. Geophys. Res.* 89:1350–1354.

Grosjean, D. and Lewis, R. 1982. Atmospheric photooxidation of methyl sulfide. *Geophys. Res. Lett.* 9:1203–1206.

Hamill, P., Kiang, C.S., and Cadle, R.D. 1977. The nucleation of H_2SO_4-H_2O solution aerosol particles in the stratosphere. *J. Aerosol Sci.* 34:150–162.

Hansen, J., Lacis, A., Rind, D., Russel, G., Stone, P., Fung, I., Ruedy, R., and Lerner, J. 1984. Climate sensitivity: analysis of feedback mechanisms, in *Climate Processes and Climate Sensitivity,* J.E. Hansen and T. Takahashi, Eds., American Geophysical Union, Washington, D.C., 130–163.

Hasselmann, K. 1991. Ocean circulation and climate change. *Tellus* 43A-B:82–103.

Heck, W.W., Taylor, O.C., Adams, R., Bingham, G., Miller, J., Preston, E., and Weinstein, L. 1982. Assessment of crop loss from ozone, *J. Air Pollut. Control Assoc.* 32:353–361.

Hegg, D.A. 1991. Particle production in clouds. *Geophys. Res. Lett.* 18:995–998.

Hegg, D.A., Hobbs, P.V., and Radke, L. F. 1980. Observations of the modification of cloud condensation nuclei in wave clouds. *J. Rech. Atmos.* 14:217–222.

Heicklen, J. 1976. *Atmospheric Chemistry,* Academic Press, New York.

Herbert, G.A., Bridgman, H.A., Schnell, R.C., Bodhaine, B.A., and Oltmans, S.J. 1987. *The Analysis of Meterological Conditions and Haze Distribution for the Second Arctic Gas and Aerosol Sampling Program (AGASP II), March–April 1986,* NOAA Technical Memorandum ERL ARL-158, Silver Springs, M.D.

Hicks, B.B., Meyers, T.P., and Baldocchi, D.D. 1988. Aerometric measurement requirements for quantifying dry deposition, in *Principles of Environmental Sampling,* L.H. Keith, Ed., American Chemical Society Professional Reference Book, Washington D.C. 297–315.

Hirschler, M.M. 1981. Man's emission of carbon dioxide into the atmosphere. *Atmos. Environ.* 15:719–727.

Hofmann, D. J. 1990. Increase in the stratospheric background sulfuric acid aerosol mass in the past 10 years. *Science* 248:996–1000.

Hofmann, D.J. and Solomon, S. 1989. Ozone destruction through heterogeneous chemistry following the eruption of El Chichón. *J. Geophys. Res.* 94:5029–5042.

Holland, H.D. 1984. *The Chemical Evolution of the Atmosphere and Oceans,* Princeton University Press, Princeton, N.J.

Horváth, L. 1983. Trend of nitrate and ammonium content of precipitation water in Hungary for the last 80 years. *Tellus* 35B:304–308.

Hough, A.M. and Johnson, C.E. 1991. Modelling the role of nitrogen oxides, hydrocarbons and carbon monoxide in the global formation of tropospheric oxidants. *Atmos. Environ.* 25A:1819–1836.

Houghton, R.A., Boone, R.D., Fruci, J.R., Hobbie, J.E., Melillo, J.M., Palm, C.A., Peterson, B.J., Shaver, G.R., Woodwell, G.M., Moore, B., Skole, D.L., and Myers, N. 1987. The flux of carbon from terrestrial ecosystems to the atmosphere in 1980 due to changes in land use: geographic distribution of the global flux. *Tellus* 39B:122–139.

Hunt, B.G. 1966. Photochemistry of ozone in a moist atmosphere. *J. Geophys. Res.* 71:1385–1398.

Husar, R.B. 1986. Emissions of sulfur dioxide and nitrogen oxides and trends for eastern North America, in *Acid Deposition Long-Term Trends*, National Academy Press, Washington D.C., 48–92.

IPCC (Intergovernmental Panel on Climate Change). 1990. *Climate Change*, J.T. Houghton, G.J. Jenkins, and J.J. Ephraums, Eds., Cambridge University Press, Cambridge.

Isaksen, I.S.A. and Hov, O. 1987. Calculation of trends in the tropospheric concentration of O_3, OH, CO, CH_4 and NO_x. *Tellus* 39B:271–285.

Isidorov, V.A., Zenkevich, I.G., and Ioffe, B.V. 1990. Volatile organic compounds in solfataric gases. *J. Atmos. Chem.* 10:329–340.

Jacob, D.J. and Hoffmann, M.R. 1983. A dynamic model of the production of H^+, NO_3^- and SO_4^{2-} in urban fog. *J. Geophys. Res.* 88:6611–6621.

Jaenicke, R. and Schütz, L. 1982. Arctic aerosols in surface air. *Idöjárás* 86:235–241.

Johnston, H.S. 1971. Reduction of stratospheric ozone by nitrogen oxide catalyst from supersonic transport. *Science* 173:517–522.

Johnston, H.S. and Podolske, J. 1978. Interpretations of stratospheric photochemistry. *Rev. Geophys. Spce Phys.* 16:491–519.

Johnston, H.S., Kiunison, D.E., and Wuebbles, D.J. 1989. Nitrogen oxides from high-altitude aircraft: an update of potential effects on ozone. *J. Geophys. Res.* 94:16351–16363.

Junge, C.E. 1962. Global ozone budget and exchange between stratosphere and troposphere. *Tellus* 14:362–377.

Junge, C.E. 1963. *Air Chemistry and Radioactivity*. Academic Press, New York.

Junge, C.E., Chagnon, C.W., and Manson, J.E. 1961. Stratospheric aerosols. *J. Meteorol.* 18:81–108.

Junge, C.E. and Jaenicke, R. 1971. New results in background aerosol studies from the Atlantic expedition of the R. V. Meteor, Spring 1969. *J. Aerosol Sci.* 2:305–314.

Kasting, J.F., Liu, S.C., and Donahue, T.M. 1979. Oxygen levels in the primitive atmosphere. *J. Geophys. Res.* 84:3097–3107.

Kauppi, P., Posch, M., Kauppi, L., and Kamari, J. 1990. Modeling soil acidification in Europe, in *The RAINS Model of Acidification. Science and Strategies in Europe*, J. Alcamo, R. Shaw, and L. Hordijk, Eds., Kluwer Academic Publishers, Dordrecht, Netherlands, 179–222.

Keeling, C.D., Bacastow, R.B., Carter, A.F., Piper, S.C., Whorf, T.P., Heimann, M., Mook, W.G., and Roeloffzen, H. 1989. A three dimensional model of atmospheric CO_2 transport based on observed winds: 1. Analysis of observational data in aspects of climate variability in the Pacific and the Western Americas, in *Geophysical Monograph 55,* D.H. Peterson, Ed., AGU, Washington, D.C., 165–236.

Khalil, M.A.K. and Rasmussen, R.A. 1989. Temporal variations of trace gases in ice cores, in *The Environmental Record in Glaciers and Ice Sheets,* H. Oeschger and C.C. Langway, Jr., Eds., John Wiley & Sons, New York, 193–205.

Khalil, M.A.K. and Rasmussen, R.A. 1990. Global increase of atmospheric molecular hydrogen. *Nature* 347:743–745.

Kondratyev, K.Ya. and Binenko, V.I. 1987. Optical properties of dirty clouds, in *Interaction between Energy Transformations and Atmospheric Phenomena,* M. Beniston and R.A. Pielke, Eds., Reidel, Dordrecht, Netherlands.

Langner, J. and Rodhe, H. 1991. A global three-dimensional model of the tropospheric sulfur cycle. *J. Atmos. Chem.* 13:225–264.

Legrand, M.R. and Delmas, R.J. 1984. The ionic balance of Antarctic snow: a 10-year detailed record. *Atmos. Environ.* 18:1867–1874.

Legrand, M.R. and Delmas, R.J. 1986. Relative contributions of tropospheric and stratospheric sources to nitrate in Antarctic snow. *Tellus* 38B:236–249.

Legrand, M.R., Delmas, R.J., and Charlson, R.J. 1988. Climate forcing implications from Vostok ice-core sulfate data. *Nature* 334:418–420.

Leighton, P.A. 1961. *Photochemistry of Air Pollution.* Academic Press, New York.

Lelieveld, J. and Crutzen, P.J. 1990. Influences of cloud photochemical processes on tropospheric ozone. *Nature* 343:227–233.

Likens, G.E., Bormann, F.H., Pierce, R.S., Eaton, J.S., and Johnson, N.M. 1977. *Biochemistry of a Forested Ecosystem,* Springer-Verlag, Berlin.

Lin, X., Trainer, M., and Liu, S.C. 1988. On the nonlinearity of the tropospheric ozone production. *J. Geophys. Res.* 93:15,879–15,888.

Lindstrom, D.R. and MacAyeal, D.R. 1989. Scandinavian, Siberian, and Arctic ocean glaciation: effect of Holocene atmospheric CO_2 variations. *Science* 245:628–631.

Lipschultz, F., Zafariou, O.C., Wofsy, S.C., McElroy, M.B., Valois, F.W., and Watson, S.W. 1981. Production of NO and N_2O by soil nitrifying bacteria. *Nature* 294:641–643.

Lodge, J.P. 1955. A study of sea-salt particles over Puerto Rico. *J. Meteor.* 12:493–499.

Logan, J.A. 1983. Nitrogen oxides in the troposhere: global and regional budgets. *J. Geophys. Res.* 88:10785–10807.

Logan, J.A., McElroy, M.B., Wofsy, S.C., and Prather, M.J. 1979. Oxidation of CS_2, and COS: sources for atmospheric SO_2. *Nature* 281:185–188.

Logan, J.A., Prather, M.J., Wofsy, S.C., and McElroy, M.B. 1981. Tropopheric chemistry: a global perspective. *J. Geophys. Res.* 86:7210–7254.

Lopez, A., Prieur, S., and Fontan, J. 1984. Study of the formation of particles from natural hydrocarbons released by vegetation, in *11th International Conference on Atmospheric Aerosols, Condensation, and Ice Nuclei. Pre-Print Volume I,* Budapest, 35–51.

Lovelock, J.P. 1988. *The Ages of Gaia. A Biography of Our Living Earth.* The Commonwealth Fund Book Program, Oxford University Press, Oxford.

Lovelock, J.P. and Margulis, L. 1974. Atmospheric homeostasis by and for the biosphere: the Gaia hypothesis. *Tellus* 26:2–10.

MacCracken, M.C. 1990. *Energy and Climate Change. Report of the DOE Multi-Laboratory Climate Change Committee.* Lewis Publishers, MI.

Manabe, S. and Wetherald, R.T. 1987. Thermal equilibrium of the atmosphere with a given distribution of relative humidity. *J. Atmos. Sci.* 24:241–259.

McElroy, M.B. 1986. Change in the natural environment of the Earth: the historical record, in *Sustainable Development of the Biosphere,* B.C. Clark and R.E. Munn, Eds., Cambridge University Press, Cambridge, 199–211.

Mészáros, A. and Vissy, K. 1974. Concentration, size distribution and chemical nature of atmospheric aerosol particles in remote oceanic areas. *J. Aerosol Sci.,* 5:101–110.

Mészáros, A., Haszpra, L., Kiss, I., Koltay, E., László, S., and Szabó, Gy. 1984. Trace element concentrations in atmospheric aerosol over Hungary, in *11th International Conference on Atmospheric Aerosols, Condensation and Ice Nuclei, Pre-Print Volume I,* Budapest, 113–117.

Mészáros, E. 1964. Répartition verticale de la concentration des particules de chlorures dans les basses couches de l'atmosphère. *J. Rech. Atmos.* 1(2^e année):1–10.

Mészáros, E. 1978. Concentration of sulfur compounds in remote continental and oceanic areas. *Atmos. Environ.* 12:699–705.

Mészáros, E. 1991a. The atmospheric aerosol, in *Atmospheric Particles and Nuclei*, G. Götz, E. Mészáros, and G. Vali, Eds., Akadémiai Kiadó, Budapest, 17–84.

Mészáros, E. 1991b. Cloud condensation nuclei, in *Atmospheric Particles and Nuclei*, G. Götz, E. Mészáros, and G. Vali, Eds., Akadémiai Kiadó, Budapest, 85–130.

Mészáros, E. 1992. Occurrence of atmospheric acidity, in *Atmospheric Acidity—Sources, Consequences and Abatement*, M. Radojevic and R.M. Harrison, Eds., Elsevier Science Publishers Ltd., in press.

Mészáros, E. and Várhelyi, G. 1982. An evaluation of the possible effect of anthropogenic sulfate particles on the precipitation ability of clouds over Europe. *Idöjárás* 86:76–81.

Mészáros, E. and Pálvölgyi, T. 1990. Daisyworld with an atmosphere. *Idöjárás* 94:339–345.

Miller, J.M. 1984. Acid rain. *Weatherwise* 37:233–239.

Miller, S.L. 1953. Production of amino acids under possible primitive Earth conditions. *Science* 117:528–529.

Mirabel, P.J. and Jaecker-Voirol, A. 1988. Binary homogeneous nucleation, in *Atmospheric Aerosols and Nucleation*, P.E. Wagner and G. Vali, Eds., Springer-Verlag, Berlin, 3–14.

Molina, M. and Rowland, F.S. 1974. Stratospheric sink for chlorofluoromethanes-chlorine atom catalysed destruction of ozone. *Nature* 249:810–812.

Molnár, A. 1990. Estimation of volatile organic compounds (VOC) emissions for Hungary. *Atmos. Environ.* 24A:2855–2860.

Moore, D.J. and Mason, B.J. 1954. The concentration, size distribution and production rate of large salt nuclei over the oceans. *Q. J. R. Meteorol. Soc.* 80:583–590.

Munn, R.E. and Rodhe, H. 1985. *Compendium of Meteorology*, WMO, Geneva.

NAS (National Academy of Sciences). 1979. *Stratospheric Ozone Depletion by Halocarbons: Chemistry and Transport*, Washington, D.C.

Nefter, A., Beer, J., Oeschger, H., Zürcher, F., and Finkel, R.C. 1985. Sulfate and nitrate concentrations in snow from South Greenland 1895–1978. *Nature* 314:611–613.

Nguyen, B.C., Bonsang, B., and Gaudry, A. 1983. The role of the ocean in the global atmospheric sulfur cycle. *J. Geophys. Res.* 88:10,903–10,914.

Nodop, K. 1986. Nitrate and sulfate wet deposition in Europe, in *Physico-Chemical Behaviour of Atmospheric Pollutants,* Reidel, Dordrecht, Netherlands, 520–528.

Novakov, T. 1984. The role of soot and primary oxidants in atmospheric chemistry. *Sci. Total Environ.* 36:1–10.

Nriagu, J.O. 1989. A global assessment of natural sources of atmospheric trace metals. *Nature* 338:47–49.

Nriagu, J.O. and Pacyna, J.M. 1988. Quantitative assessment of worldwide contamination of air, water and soil by trace metals. *Nature* 333:134–139.

Ogren, J.A., Groblicki, P.J., and Charlson, R.J. 1984. Measurement of the removal rate of elemental carbon from the atmosphere. *Sci. Total Environ.* 36:329–338.

Owen, T., Cess, R.D., and Ramanathan, V. 1979. Enhanced CO_2 greenhouse to compensate for reduced solar luminosity on early Earth. *Nature* 277:640–642.

Pacyna, J.M. 1981. *Emission Factors of Trace Metals from Coal-Fired Power Plants.* NILU Teknisk rapport nr: 14/81, Lillestrom.

Pacyna, J.M., Bartonova, A., Cornille, P., and Maenhaut, W. 1989. Modelling of long-range transport of trace elements. A case study. *Atmos. Environ.* 23:107–114.

Pacyna, J.M., Münch, J., Alcamo, J., and Anderberg, S. 1991. Emission trends for heavy metals in Europe, in *Proc. Int. Conference on Heavy Metals in the Environment,* Glasgow.

Pearman, G.I., Hyson, P., and Fraser, P.J. 1983. The global distribution of atmospheric carbon dioxide: 1. Aspects of observations and modeling. *J. Geophys. Res.* 88:3581–3590.

Penkett, S.A. 1989. Ultraviolet levels down not up. *Nature* 341:283–284.

Penkett, S.A. 1991. International Conference on the Generation of Oxidants on Regional and Global Scales, Norwich, 3–7 July 1989: an overview. *Atmos. Environ.* 25A:1735–1737.

Penkett, S.A., Jones, B.M.R., Brice, K.A., and Eggleton, A.E.J. 1979. The importance of ozone and hydrogen peroxide in oxidising sulphur dioxide in cloud and rainwater. *Atmos. Environ.* 13:123–137.

Penkett, S.A., Derwent, R.G., Fabian, P., Borchers, R., and Schmidt, U. 1980. Methyl chloride in the stratosphere. *Nature* 283:58–60.

Penner, J.E., Atherton, C.S., Dignon, J., Ghan, S.J., Walton, J.J., and Hameed, S. 1991. Tropospheric nitrogen: a three-dimensional study of sources, distributions, and deposition. *J. Geophys. Res.* 96:959–990.

Petersen, G., Weber, H., and Grassl, H. 1989. Modelling the atmospheric transport of trace metals from Europe to the North Sea and the Baltic Sea, in *Control and Fate of Atmospheric Trace Metals*, J.M. Pacyna and B. Ottar, Eds., Kluwer Academic Publishers, Dordrecht, Netherlands, 57–84.

Podzimek, J. 1984. Size spectra of bubbles in the foam patches and of sea salt nuclei over the surf zone. *Tellus* 36B:192–202.

Potter, G.L. and Cess, R.D. 1984. Background tropospheric aerosols: incorporation within a statistical-dynamical model. *J. Geophys. Res.* 89:9521–9526.

Prather, M.J. and Watson, R.T. 1990. Stratospheric ozone depletion and future levels of atmospheric chlorine and bromine. *Nature* 344:729–734.

Prospero, J.M. 1968. Atmospheric dust studies on Barbados. *Bull. Am. Meteorol. Soc.* 49:645–652.

Prospero, J.M. 1984. Aerosol particles, in *Global Tropospheric Chemistry*. National Academy Press, Washington, D.C., 136–140.

Puxbaum, H. 1991. Metal compounds in the atmosphere, in *Metals and Their Compounds in the Environment*, E. Merian, Ed., VCH Verlagsgesellschaft mbH, Weinheim, Germany, 257–286.

Raatz, W.E. 1985. Meteorological conditions over Eurasia and the Arctic contributing to the March 1983 Arctic haze episode. *Atmos. Environ.* 19:2121–2126.

Rahn, K.A. 1976. *The Chemical Composition of the Atmospheric Aerosol*. University of Rhode Island, Technical Report, Kingston, RI.

Rahn, K.A. 1981. Relative importances of North America and Eurasia as sources of Arctic aerosol. *Atmos. Environ.* 15:1447–1455.

Ramanathan, V. 1988. The greenhouse theory of climate change: a test by an inadvertent global experiment. *Science* 240:293–299.

Ramanathan, V., Cess, R.D., Harrison, E.F., Minnis, P., Barkstrom, B.R., Ahmad, E., and Hartmann, D. 1989. Cloud-radiative forcing and climate: results from the Earth Radiation Budget Experiment. *Science* 243:57–63.

Rasmussen, R.A., Khalil, M.A.K., and Hoyt, S.D. 1982. The oceanic source of carbonyl sulfide (OCS). *Atmos. Environ.* 16:1591–1594.

Ratner, M.I. and Walker, J.C.G. 1972. Atmospheric ozone and the history of life. *J. Atmos. Sci.* 29:803–808.

Rodhe, H. and Granat, L. 1984. An evaluation of sulfate in European precipitation 1955–1982. *Atmos. Environ.* 18:2627–2639.

Rodriguez, J.M., Ko, M.K.W., and Sze, N.D. 1991. Role of heterogeneous conversion of N_2O_5 on sulfate aerosols in global ozone losses. *Nature* 352:134–137.

Rosen, H., Novakov, T., and Bodhaine, B.A. 1981. Soot in the Arctic. *Atmos. Environ.* 15:1371–1374.

Rosen, H., Hansen, A.D.A., and Novakov, T. 1984. Role of graphitic carbon particles in radiative transfer in the Arctic Haze. *Sci. Total Environ.* 36:103–110.

Rotty, R.M. 1987. A look at 1983 CO_2 emissions from fossil fuels (with preliminary data for 1984). *Tellus* 39B:203–208.

Russel, P.B. and Hamill, P. 1984. Spatial variation of stratospheric aerosol acidity and model refractive index: implications of recent results. *J. Atmos. Sci.* 41:1781–1790.

Ryaboshapko, A.G. 1983. The atmospheric sulfur cycle, in *The Global Biogeochemical Sulfur Cycle,* M.V. Ivanov and J.R. Freney, Eds., John Wiley & Sons, Chichester, U.K., 203-296.

Schidlowski, M. 1978. A model for the evolution of photosynthetic oxygen. *Pure Appl. Geophys.* 116:234–238.

Schlesinger, M.E. and Zhao, Z.C. 1989. Seasonal climatic changes induced by doubled CO_2 as simulated by the OSU Atmospheric GCM/Mixed-layer Ocean Model. *J. Climate* 2:459–495.

Schneider, B., Tindale, N.W., and Duce, R.A. 1990. Dry deposition of Asian mineral dust over the central North Pacific. *J. Geophys. Res.* 95:9873–9878.

Schneider, S.H. and Thompson, S.L. 1988. Simulating the climatic effects of nuclear war. *Nature* 333:221–227.

Schoeberl, M.R., Stolarski, R.S., and Krueger, A.J. 1989. The 1988 Antarctic ozone depletion: comparison with previous year depletions. *Geophys. Res. Lett.* 16:377–380.

Schwartz, S.E. 1988. Are global cloud albedo and climate controlled by marine phytoplankton? *Nature* 336:441–445.

Seiler, W., Giehl, H., Brunke E.G., and Halliday, E. 1984. The seasonality of CO abundance in the Southern Hemisphere. *Tellus* 36B:219–231.

Seiler, W. 1974. The cycle of atmospheric CO. *Tellus* 26:116–135.

Seiler, W. and Conrad, R. 1987. Contribution of tropical ecosystems to the global budgets of trace gases especially CH_4, H_2, CO, and N_2O, in *Geophysiology of Amazonia*, R. Dickinson, Ed., John Wiley & Sons, New York, 133–162.

Shaw, G.E. 1987. Aerosols as climate regulators: a climate-biosphere linkage? *Atmos. Environ.* 21:985–986.

Sheppard, J.C., Westberg, H., Hopper, J.F., Ganesan, K., and Zimmerman, P. 1982. Inventory of global methane sources and their productions rates. *J. Geophys. Res.* 87:1305–1312.

Siegenthaler, U. and Oeschger, H. 1987. Biospheric CO_2 emissions during the past 200 years reconstructed by deconvolution of ice core data. *Tellus* 39B:140–154.

Sigg, A. and Nefter, A. 1991. Evidence for a 50% increase in H_2O_2 over the past 200 years from a Greenland ice core. *Nature* 351:557–559.

Simpson, D. 1991. *Long Period Modelling of Photochemical Oxidants in Europe. Calculations for April-September 1985, April-October 1989*, EMEP/MSC-W, Report 2/91, Oslo.

Singh, H.B., Salas, L.J., Shigeishi, H., and Scribner, E. 1979. Atmospheric halocarbons, hydrocarbons and sulfur hexafluoride: global distributions, sources and sinks. *Science* 203:899–903.

Singh, H.B. and Salas, L.J. 1983. Peroxyacetyl nitrate in the free troposphere. *Nature* 302:326–328.

SMIC. 1971. *Inadvertent Climate Modification*. MIT Press, Cambridge, MA.

Smith, F.B. 1991. Deposition processes for airborne pollutants. *Meteorol. Mag.* 120:173–182.

Solomon, S., Mount, G.H., Sanders, R.W., Jakoubek, R.O., and Schmeltekopf, A.L. 1988. Observations of the nighttime abundance of OClO in the winter stratosphere above Thule, Greenland. *Science* 242:550–555.

Staehelin, J. and Schmid, W. 1991. Trend analysis of tropospheric ozone concentrations utilizing the 20-year data set of ozone balloon soundings over Payerne (Switzerland). *Atmos. Environ.* 25A:1739–1750.

Stelson, A.W., Friedlander, S.K., and Seinfeld, J.H. 1979. A note on the equilibrium relationship between ammonia and nitric acid and particulate ammonium nitrate. *Atmos. Environ.* 13:369–371.

Stensland, G.J., Whelpdale, D.M., and Oehlert, G. 1986. Precipitation chemistry, in *Acid Deposition Long-Term Trends*, National Academy Press, Washington, D.C., 128–199.

Stephenson, J.A.E. and Scourfield, M.W.J. 1991. Importance of energetic solar protons in ozone depletion. *Nature* 352:137–139.

Stolarski, R.S. and Cicerone, R.J. 1974. Stratospheric chlorine: a possible sink for ozone. *Can. J. Chem.* 52:1610.

Stolarski, R.S., Bloomfield, P., McPeters, R.D., and Herman, J.R. 1991. Total ozone trends deduced from Nimbus 7 TOMS data. *Geophys. Res. Lett.* 18:1015–1018.

Swedish Ministry of Agriculture. 1982. *Acidification Today and Tomorrow.* Tryckeri A.B., Risbergs, Uddevalla.

Tans, P.P., Fung, I.Y., and Takahashi, T. 1990. Observational constraints on the global atmospheric CO_2 budget. *Science* 247:1431–1438.

Thompson, A.M., Huntley, M.A., and Stewart, R.W. 1991. Perturbations to tropospheric oxidants, 1985–2035: 2. Calculations of hydrogen peroxide in chemically coherent regions. *Atmos. Environ.* 25A:1837–1850.

Titus, J.G. and Seidel, S. 1988. Overview of the effects of changing the atmosphere, in *Effects of Changes in Stratospheric Ozone and Global Climate Volume 2,* J.G. Titus, Ed., U.S. Environmental Protection Agency, 3–19.

Turco, R.P. 1982. Models of stratospheric aerosols and dust, in *The Stratospheric Aerosol Layer,* R.C. Whitten, Ed., Springer-Verlag, Berlin, 93–119.

Twomey, S. 1955. The distribution of sea-salt nuclei in air over land. *J. Meteor.* 12:81–86.

Twomey, S. 1977. *Atmospheric Aerosols.* Elsevier, Amsterdam.

Twomey, S. 1991. Aerosols, clouds and radiation. *Atmos. Environ.* 25A:2435–2442.

Vali, G. 1991. Nucleation of ice, in *Atmospheric Particles and Nuclei,* G. Götz, E. Mészáros, and G. Vali, Eds., Akadémiai Kiadó, Budapest, 131–192.

Voldner, E.C., Barrie, L.A., and Sirois, A. 1986. A literature review of dry deposition of oxides of sulphur and nitrogen with emphasis on long-range transport modelling in North America. *Atmos. Environ.* 20:2101–2123.

Volz, A.D. and Kley, D. 1988. Evaluation of the Montsouris ozone measurements made in the nineteenth century. *Nature* 332:240–242.

Vong, R.J., Hansson, H.-C., Covert, D.S., and Charlson, R.J. 1988. Acid rain: simultaneous observations of a natural marine background and its acidic sulfate aerosol precursor. *Geophys. Res. Lett.* 15:338–341.

Walker, J.C.G. 1974. Stability of atmospheric oxygen. *Am. J. Sci.* 274:193–214.

Walker, J.C.G. 1977. *Evolution of the Atmosphere,* Macmillian Publishing, New York.

Walker, J.C.G., Hays, P.B., and Kasting, J.F. 1981. A negative feedback mechanism for the long-term stabilization of the Earth's surface temperature. *J. Geophys. Res.* 86:9776–9782.

Warneck, P. 1988. *Chemistry of the Natural Atmosphere,* R. Dmowska and J.R. Holton, Eds., Academic Press, San Diego.

Washington, W.M. and Meehl, G.A. 1984. Climate sensitivity due to increased CO_2: experiments with coupled atmosphere and ocean general circulation model. *Climate Dynamics* 4:1–38.

Watson, A.J. and Lovelock, J.E. 1983. Biological homeostasis of the global environment: the parable of Daisyworld. *Tellus* 35B:284–289.

Watson, R.T. 1989. Stratospheric ozone depletion: Antarctic processes, in *Ozone Depletion, Greenhouse Gases, and Climate Change,* National Academy Press, Washington, D.C., 19–32.

Weizsäcker, E.U. 1990. Prices should tell the ecological truth, in *Report of the Conference on Sustainable Development Science and Policy,* Bergen, Norway, 523–535.

Went, F.W. 1966. On the nature of Aitken condensation nuclei. *Tellus* 18:549–556.

Whitby, K.T. 1978. The physical characteristics of sulfur aerosols. *Atmos. Environ.* 12:135–159.

Wigley, T.M.L. 1989. Possible climate change due to SO_2-derived cloud condensation nuclei. *Nature* 339:365–367.

WMO. 1976. WMO statement on modification on the ozone layer due to man's activities and some possible geophysical consequences. *WMO Bull.* 25:59–63.

Wofsy, S.C. and Logan, J.A. 1982. Recent developments in stratospheric photochemistry, National Research Council, in *Causes and Effects of Stratospheric Ozone Reduction: An Update,* National Academy Press, Washington, D.C., 167–205.

Woodcock, A.H. 1953. Salt nuclei in marine air as a function of altitude and wind force. *J. Meteor.* 10:362–371.

Woodwell, G.M. 1989. Biotic causes and effects of the disruption of the global carbon cycle, in *The Challenge of Global Warming,* D.E. Abrahamson, Ed., Island Press, Washington, D.C., 71–81.

Zardini, D., Raynaud, D., Scharffe, D., and Seiler, W. 1989. N_2O measurements of air extracted from Antarctic ice cores: implication on atmospheric N_2O back to the last glacial-interglacial transition. *J. Atmos. Chem.* 8:189–201.

Zeng, X., Pielke, R.A., and Eykholt, R. 1990. Chaos in daisyworld. *Tellus* 42B:309–318.

Zimmerman, P.R., Chatfield, R.B., Fishman, J., Crutzen, P.J., and Hanst, P.L. 1978. Estimates on the production of CO and H_2 from the oxidation of hydrocarbon emissions from vegetation. *Geophys. Res. Lett.* 5:679–682.

Zoller, W.H., Cunningham, W.C., and Duce, R.A. 1979. Trends and composition of atmospheric aerosols from South Pole station, Antarctica. *WMO Special Environ. Report No. 14,* Geneva, 245–258.

Index